Reading Popular Newtonianism

Laura Miller

Reading Popular Newtonianism

Print, the *Principia*, and the Dissemination of Newtonian Science

University of Virginia Press
Charlottesville & London

University of Virginia Press

Printed in the United States of America on acid-free paper

First published 2018

ISBN 978-0-8139-4125-7 (cloth)
ISBN 978-0-8139-4126-4 (ebook)

9 8 7 6 5 4 3 2 1

Library of Congress Cataloging-in-Publication Data is available for this title.

Cover art: From the title pages of Isaac Newton's *Philosophiae Naturalis Principia Mathematica* (1687) and *Opticks* (1704)

For my sons and my mother

What wonder thence that his devotion swell'd
Responsive to his knowledge? For could he,
Whose piercing mental eye diffusive saw
The finish'd university of things
In all its order, magnitude, and parts,
Forbear incessant to adore that Power
Who fills, sustains, and actuates the whole?

—James Thomson, "A Poem Sacred to the Memory of Sir Isaac Newton"

Contents

Illustrations

Acknowledgments

This project originated in an experience of reading popular Newtonianism: my first encounter with Francesco Algarotti's *Il Newtonianismo per le dame* (1737). I was at the Dickens Universe in the UC Santa Cruz dorms, crouched in a corner of my room trying to access the internet, paging through everything I could find on ECCO related to Newton, about whose *Opticks* I had written in Anita Guerrini's seminar that spring. And there Algarotti's book was, in Elizabeth Carter's 1739 English translation. Here was a work about Newton marketed to women—one that ostensibly existed because women needed to be protected from the indelicacy of math. Its attempts to teach Newtonian science through titillating prose and its multiple editions throughout the eighteenth century gestured to a diverse and enduring readership that intrigued me. This furtive inquiry at a Victorian conference bred in me a desire to investigate popular Newtonianism further, and has resulted in the book you are reading.

I would first like to thank my mentors from the University of California, Santa Barbara. William B. Warner, E. Heckendorn Cook, and Anita Guerrini helped to develop this project in its earliest stages and have shaped me as a scholar throughout my career. Elsewhere at UCSB, Patricia Fumerton and the English Broadside Ballad Archive—via the NEH—kept me employed during summers of writing. A research seminar led by Ann Bermingham at UCSB's Interdisciplinary Humanities Center came along at a crucial time for my thinking about Newtonian popularization and material culture.

My colleagues at the University of West Georgia have shown enthusiastic support for this project, including Meg Pearson and Margaret Mitchell. My friend and office neighbor Patrick Erben has been generous and patient in responding to my many early Americanist questions. The grace and productivity of my department as a whole inspire me on a regular basis. Graduate students Jenna Harte and Benjamin Fuller offered helpful assistance with footnotes and proofreading.

Many, many granola bars were eaten in haste outside the doors of research libraries during the writing of this book. This research was supported by generous grants from UCSB, UWG, the William Andrews Clark Memorial Library, and the Houghton Library. Visits to other archives were essential to the project's success, including the British Library, the American Antiquarian Society, the Wren Library at Trinity College, Cambridge, the Bodleian Library, the Huntington Library, the *Encyclopaedia Britannica* archives, the New York Society Library, and Columbia University's Rare Book and Manuscript Library.

Portions of chapter 4 originally appeared in *Studies in Eighteenth-Century Culture* 42 (2013). I am grateful for *SECC*'s permission to reproduce them here. The Wellcome Library, the Huntington Library, and the National Portrait Gallery have been speedy and helpful with imaging and permissions.

Finding a group of kind, brilliant, and generous people in eighteenth-century studies to get to know over the years has been a great pleasure. So many scholars have taken the time to read and comment on my work at conferences, as well as offer friendship and support in private life. Because of my exceptional colleagues in eighteenth-century studies, I can feel that the world is a smaller, more affectionate place than it actually is. To that end, I offer special thanks to, among many others, Emily Friedman, James Horowitz, Jason Pearl, Carrie Shanafelt, Rivka Swenson, and Brett D. Wilson. I would also like to thank Danielle Spratt and Courtney Weiss Smith, who offered feedback on this manuscript at critical stages in the writing process. I must also recognize the inspiration of the AHRC Community Libraries network as a body of generous colleagues whom I am fortunate to know, and who were instrumental in helping me with my fifth chapter—in particular Mark Towsey, Kyle Roberts, and Rob Koehler.

The University of Virginia Press has been ideal to work with, especially Angie Hogan, Ellen Satrom, and copyeditor Colleen Romick Clark. I am indebted to the anonymous readers for the press, whose excellent feedback has helped make this a better book.

I wish also to thank my friends and their valiant attempts to help me live a normal life. Those I am likely to message immediately when crisis strikes—Omi Miri and Jason Spicer—deserve extra commendation for their wisdom, practicality, and boundless wit.

I give special thanks to Patrick Ludolph for his support of me during the years of research and writing that went into this book, and for being such an excellent father to George and Ned. And I am grateful to my father,

Roy Miller, for his cheerful encouragement and support of my writing and scholarship.

Reading Popular Newtonianism is dedicated to my late mother, Virginia Miller, and my beautiful, loving sons, George and Ned Ludolph, who never met her. I can at least bring them together here in a medium that flattens out time, distance, and mortality: the book.

Reading Popular Newtonianism

Introduction

Reading Popular Newtonianism: Print, the "Principia," and the Dissemination of Newtonian Science analyzes the reception of Sir Isaac Newton's (1642–1727) works in a context framed by authorship, print, editorial practices, and reading. Although Newton's first publications were journal articles that dated from the early 1670s, Newton's print reception truly begins with the publication of the landmark monograph *Philosophiae Naturalis Principia Mathematica* (*Principia*) in 1687.[1] Newton's manipulation of printing elements, such as front matter, diagrams, and the formal organization of its narrative, lent the *Principia* a superficial intelligibility that promoted the understanding of its concepts even if its readers could not fully grasp the mathematical principles of natural philosophy that the *Principia* detailed.[2] The broader cultural relevance of the *Principia*'s foundational concepts became evident as these ideas found homes in scientific and cultural discourses. Many channels effected this dissemination, including written responses, lectures, public experiments, published "popularizations," and library reading. Indeed, concepts like attraction and extrapolation—fundamental to the *Principia*—suffuse Newtonian popularizations.[3] The foregrounding of these concepts facilitated the dissemination of other cultural markers—such as the connections between attraction and desire, debates about masculinity, and the tension between individuals and communities that emerge in imaginative literature of the period. Such grand effects, this book contends, are rooted in the publication practices associated with the *Principia*. Newton's vision of print as a medium in which concrete metaphor and mathematical abstraction formed a unified whole reoriented Britain and the West toward the macrocosm of the universe itself.

The term "Newtonianism" refers to the dissemination of Newton's ideas in Britain, continental Europe, and the Americas. Popular Newtonianism is the transmission of Newton's ideas to Western culture at large through an assortment of printed works and live demonstrations.[4] In this volume, I focus on the print reception of popular Newtonianism—I read popular Newtonianism—as a phenomenon whose roots can be traced to the *Principia*. Whether people read nonfiction "popularizations," poetry, or writings on aesthetics that incorporated Newtonian principles, print facilitated the mainstreaming of Newton's ideas.

Beyond its consideration of print and popular Newtonianism, *Reading Popular Newtonianism* crystallizes our understanding of eighteenth-century authorship, publication, and readership, while illuminating the permeability of disciplinary boundaries in the arts and sciences, and the correspondingly diffuse dissemination of scientific ideas in the eighteenth century. The story of Newton, popularization, and print furthers our understanding of each. Patricia Fara identifies the problems of circularity in studying the popularization of Newton: because so much of Newton is filtered through myth, it is difficult to know which narratives of Newton's origins were constructed after he became well known.[5] Fara writes that "in the absence of any formal structures for scientific education, people learned about Newton from chatty journal articles or reading poetry, then an enormously popular genre for transmitting ideas."[6] Readers also, I show, learned much about Newton from the *Principia* itself and its paratexts, as well as informal educational publications like encyclopedias, poetry, and books on the "sister arts." Newton's ideas were disseminated in an increasingly diversifying print marketplace. The works of popular Newtonianism produced during this period reflect an imagined readership to which the books were marketed, a commercially grounded assumption that did not fully align with the actual reception of these texts.

This book begins by evaluating Newton's use of print through my analysis of his reading encounters with printed works, followed by my analysis of the links between print and the *Principia*. The three chapters that follow the chapter on the *Principia* focus on the interpretive communities addressed by publication—communities imagined through poetry, prose, and library books.[7] The book is organized chronologically, beginning with Newton's initial publications on optics from the early 1670s, and concluding with the end of the New York Society Library's first charging ledger in 1792. The narrative also progresses outward, from Newton's individual orientation to print to larger, decentered communities. Here I analyze the tension between "popularizers" and implied readers, as well as emerging publics composed of real readers whose preferences we can trace and interpret, as in my case-study approach to NYSL borrowers.[8]

Scholars who work on the connections between eighteenth-century science and eighteenth-century literature often focus on the comparable theatricality of experimental philosophy—where experiments were staged and witnessed—and actual theatrical performances. This method of approaching the intersections of literature and science often requires putting pressure on concepts like virtual witnessing. Sociohistorical approaches to the history of science like these have found the concept of virtuality fruitful to analyze.[9] In virtual witnessing, experiments are described in prose such that the reader becomes a virtual witness to a live, performed experiment. Literary critics have done innovative work responding to these concepts in publications that connect theatricality, literature, and experimental philosophy.[10] Scholars like Jonathan Kramnick, Helen Thompson, and Courtney Weiss Smith, who emphasize the philosophy of natural philosophy through work on empiricism, Lockean subjectivity, and new materialisms, produce essential scholarship at the borders of that discipline.[11]

The *Principia*'s narrative differed from the style of reading, and invocation of witnessing, present in the *Philosophical Transactions*—a style that has received much attention in interdisciplinary humanities scholarship. Much as Donna Haraway problematizes Robert Boyle's work and the science of the late seventeenth-century Royal Society—in particular the literary innovation of modest witnessing—Newton's writing was likewise formed deliberately.[12] The tight construction of a printed prose book like the *Principia,* following the circulation of a series of shorter printed texts, evokes the bricolage of the novel's development in Britain. The modest witness is, to Haraway, "suspicious, implicated, knowing, ignorant, worried, and hopeful."[13] With Newton's self-presentation in the *Principia,* witnessing works differently: uncertainty gives way to mathematical solutions, and lingering doubts are attributable to teleological arguments. The *Principia* illuminates the way we understand print and the figure of the modest witness—especially through resistance.

Newtonian natural philosophy posed representational challenges to eighteenth-century culture: theories like Newton's that were more abstract and that were impossible to witness literally or virtually were still culturally significant and had to be represented in popular culture somehow—through varying modes of representation. The introduction to Francesco Algarotti's *Il Newtonianismo per le dame* (the subject of chapter 4) claimed that reading its dialogues was the equivalent of spending a pleasant night at the theater. This comparison aimed to convince Algarotti's audience that the text would be entertaining and not a dry tome. Some texts, like the poems I explore in chapter 3, used the abstraction of Newtonian natural philosophy to inspire further poetic abstraction. Other eighteenth-century writers like George Horne, later bishop of Norwich, satirized the abstract nature of Newtonian natural philosophy, finding it to be incompatible with theatricality. In a letter to an unknown friend in 1749, shortly after half-term at Oxford, Horne imagined an unusual theatrical performance:

> There is a dramatical performance, which will soon make its appearance in the world, and 'tis to be hoped will meet with a favorable reception from the present learned philosophical age—the title is not yet determined upon; but a scene or two may give you some idea of the piece, and the design of it. . . . The curtain draws up and discovers—Infinite Extension—to him enter a Vacuum . . . and the vacuum after a proper time proves with child of a Plenum, and is brought to bed with dreadful throes and most acute pains, as you may well suppose. . . . The Epitholamium is sung by a Chorus of Occult Qualities. Soon after a private figure walks across the stage and disappears; upon which an Occult Quality tells the audience very coolly, that he is gone on in infinitum—just to take a turn or so before supper.[14]

Horne refers to a "learned philosophical age" composed of people who read about concepts that are unrepresentable on the stage, compared to the theatricality of experimental culture. This challenge of representation had been satirized as early as Thomas Shadwell's *The Virtuoso* (1676), where the eponymous character, Sir Nicholas Gimcrack, belly-down on a table, imitates the swimming of a frog in order to learn to swim, an event that seems far more reasonable than taking a turn "*in infinitum*" just before supper.

Al Coppola's recent book on eighteenth-century experimental philosophy and theatrical culture offers a full examination of the spectacle that accompanied eighteenth-century scientific practice. Horne's letter links well with Coppola's reading of Aphra Behn's *The Emperor of the Moon* (1687) and Nehemiah Grew's *Museum Regalis Societatis* (1681) insofar as both works advocate for "antispectacular epistemologies"; Behn's play "teach[es] a rational mindfulness to an entire society overrun with enthusiastic virtuosi."[15] Writing fifty years later, Horne criticizes the enthusiastic philosophers who appear to be disconnected from material reality through the vehicle of a failed spectacle. Spectacle, while still important to experimental philosophy and demonstration, was not the only way that natural philosophy could be interpreted. Four years after this letter was written, Horne published *A Fair, Candid, and Impartial State of the Case between Sir Isaac Newton and Mr. Hutchinson* (1753). Its seventy-six pages advocated for John Hutchinson's biblically founded natural philosophy instead of Newton's. Horne's skewering of the abstraction of Newtonian natural philosophy does not surprise, as he had lamented the disconnectedness of knowledge production from material reality.

Later in the letter, Horne incorporates Newton into this scientific spectacle: "the curtain rises to the sound of a [hundred and twenty] musicians playing in the exhausted receiver of an air-pump. At the upper end of Infinity appears a stately throne hung upon nothing, supported by a centripetal and centrifugal force—on it sits the ghost of Sr Isaac Newton, Knt. surrounded by a numerous offspring—Gravity, Attraction, Projection, Elasticity, Magnetism, and Electricity, in their proper habits and attitudes."[16] Horne again implies that qualities related to science cannot be represented in the register of performance: in this case, abstract ideas found in conversation and print are paired with instruments, like the air-pump, that are tied to experimental science. Considered together, these concepts of varying abstraction are juxtaposed awkwardly, replacing the stock characters of a drama with "their proper habits and attitudes." These concepts also resemble keywords for the reception of science in the eighteenth century, including gravity and attraction, which were essential to Newtonian science.[17] One point of Horne's satire here is that abstract concepts are impossible to interpret consistently, especially in the visual register of performance. In contrast to virtual witnessing, in which texts attempted to render the scene of experiment for the reader, Horne responds to a culture in which the reader is never intended to imagine he or she occupies the experimenter's place. Newton, described as a ghost occupying "a stately throne hung upon nothing," sits at the center of the action. Instead of being the father of an offspring from a physical relationship, as the vacuum and plenum had been, here, by parthenogenesis, Newton fathers several visually unrepresented qualities. A throne becomes all that is needed to imagine Newton himself.

This vividly imagined scene—impossible, satirical, and filled with images of faulty reproduction—reflects a defined shift in the public face of science, a shift that was reoriented around Newtonian natural philosophy in the early eighteenth century. This faulty reproduction can also be tied with print.[18] Science became linked to media of abstract dissemination such as printed texts, conversations, and interiority, and could now occupy a visual register accompanying both the polychromatic spectacle of experiment and the intimate, black-and-white medium of print. Examples like Horne's show that abstract concepts—like those in the *Principia*—needed print to be assimilated into popular culture. In some cases, like Francesco Algarotti's, theatricality would be used as a framework to help structure popular Newtonian works: even if abstract concepts could not be "performed" onstage, theatrical dialogues helped readers to understand Newtonian science.

The *Principia* helped lay the foundation for what we now understand to be scientific disciplines, including physics, astronomy, and engineering.[19] My study of the *Principia* reveals that Newton's book responded to extant print practices and extant disciplinary standards of truth, using those precepts to create a template that ultimately facilitated his popularization.[20] Beyond its contributions to both mathematics and natural philosophy, Newton's *Principia* argued that publication as a monograph was more significant, enduring, and even devout, than publication in journals like the *Philosophical Transactions of the Royal Society*, the more frequently used publication venue of Newton's late seventeenth-century British peers.[21] Newton structured the *Principia* to claim undisputedly his intellectual priority. The medium of the print monograph was the perfect place for Newton to build an argument at length, beginning with basic principles and small simulations of motion and extrapolating outward to the proof of universal gravitation. In print, the *Principia* demonstrated mathematically that the laws governing motion, attraction, and gravitation were universal on earth and across the universe.[22] This kind of sustained argument, with its incremental, interconnected proofs, could not be accomplished in a journal article. Newton used the monograph's paratexts—including its introduction and other front matter, even a dedicatory poem by Edmond Halley—to bolster his argument and distinguish his work from that of other mathematicians and natural philosophers.[23] I center my analysis on visual, networked, and technological representations of Newton's work through evaluating print, editing practices, and reception history. By beginning with the publication of the *Principia*, I show that the entry of Newton's scientific ideas into eighteenth-century culture is more diffuse than dating Newtonian popularization to the *Opticks* (1704), or to Newton's death in 1727. Newton—who had once written to Robert Hooke while the two were engaged in a publication dispute that "what's done before many witnesses is seldom without some further concern th[a]n that for truth"—fully understood the potential for corruption and misreading in experimental science as well as in print.

In 1946 Marjorie Hope Nicolson's *Newton Demands the Muse: Newton's Opticks and the Eighteenth Century Poets* cemented the reception history of Newton as connected to the 1704 publication of the more-accessible, English-language *Opticks*. Many scholars have followed her lead and have worked productively to link Newton's popularization with the *Opticks*.[24] Book chapters and articles, including the influential work of Robert Iliffe, have looked at the disparities between the otherworldly properties of Newton's publications, in which his authority elevated him above mere mortals, and his actual mortal status. Iliffe has identified the most connections between the *Principia* and print history thus far, and this book continues that conversation.[25] Adrian Johns's *The Nature of the Book* famously ends with the publication of the *Principia*, where my study begins. Newton indeed "knew how to play the stationer and win," as Johns asserts.[26] No critics have accounted sufficiently for the role of the *Principia* in long eighteenth-century print culture.

Many of the works I am writing about in this text—including encyclopedias, poems, and a work written "for the ladies"—are not scientifically innovative. However, from my perspective as someone who traces print history, these works are exceptionally relevant to eighteenth-century readers' reception of Newtonian ideas. While many scholars of print and book history have been rightly cautious not to overattribute social developments to the medium of print, the book you are about to read shows that print helped determine the ways information about figures like Newton was managed, circulated, and interpreted. *Reading Popular Newtonianism*, then, also suggests a play on words: more than a study of reading, this book also contends that we benefit from interpreting Newton's writing and related texts as material texts in order to understand how his ideas were disseminated in a marketplace of readers with varying levels of interest and expertise.

Works like Francesco Algarotti's *Newtonianism for Ladies* (1737, 1739) and Henry Pemberton's *A View of Sir Isaac Newton's Natural Philosophy* (1728) are thought to have popularized scientific concepts in Britain and on the Continent by translating complex ideas to a broader audience of nonspecialists. However, the printing of a book is not the same as the diffuse dissemination of an idea, which is why we need to look at the reception of books as well as their publication.[27] Just because a book was printed does not mean that people read it or that it was influential: only its afterlife can show us how it was received, whether it is through the demand for multiple editions, explicit reader responses to publication, or through library circulation records. The publication of *Newtonianism for Ladies* did not mean that only ladies engaged with it, nor does it mean that women only read works with "ladies" in the title. Texts beyond popular science titles—including literature, writings on aesthetics, and even encyclopedias—are also relevant to studying how popular science was received. Books that explained Newton's ideas acted as part of a network, alongside Newton's published work, its implied readership, and the orientation of subsequent and diverse publications to nonspecialist readers. To understand popular Newtonianism necessitates a return to Newton's publishing habits and practices in order to understand the trajectory of Newton's diverse critical and popular reception in print.

One ambitious task this book accomplishes is to bring together the history of science and literary history while also working in the fields of print history and critical bibliography. I interpret literature and science through the commonality of print and readership—two topics that are essential to both disciplines. Manushag N. Powell's book on the "eidolons" of periodical literature authors has helped me to frame what I call Newton's "print persona."[28] Robert Iliffe also considers "the author as idol" in a famous essay on Newton, so in a manner of speaking my book links the historian of science who has analyzed the idolatry of Newton (Iliffe) to the literary and print historian's (Powell's) careful attention to the printed medium as a vehicle for understanding authorship. Sarah Dry's book on Newton's manuscript history unites the history of Newton scholarship on manuscripts with information history.[29] Other scholars have likewise emphasized the importance of print to Newton's own career, especially the publication of the *Principia,* so it is clearly an important topic to interrogate at length.[30] Writers on adjacent topics have been exceptionally helpful as I have organized this book. Nina Gelbart's scholarly edition of Fontenelle's *Entretriens sur la pluralité des mondes* offers an excellent introduction to late seventeenth-century popular science that is contemporary to the *Principia.* Jeff Loveland's work on encyclopedias dialogs print history, popular texts, and the history of science in ways that have helped my methods here, especially in my analysis of the second edition of the *Encyclopaedia Britannica.* Shelley Costa's research on gender and the *Ladies' Diary* and Robin Rider's scholarship on the print history of eighteenth-century mathematical textbooks also inform my scholarship here.[31]

Niccolò Guicciardini has clarified the changing mathematical relevance of Newton's methods across the eighteenth century. Guicciardini's *Reading the "Principia"* focuses on the period spanning 1687–1736 and engages directly with several readers, including Newton himself, Huygens, and Leibniz. Guicciardini examines the *Principia*'s readership among late seventeenth- and early eighteenth-century experts.[32] One important distinction Guicciardini makes is to follow the separate practices of the British and the Continental readers of Newton. These practices were eventually integrated, in terms of calculus especially, by Leonhard Euler in 1736 in *Mechanica* and became standard later in the century. For experts in the practice of mathematics "after Euler[,] the *Principia*'s mathematical methods belong definitely to what is past and obsolete."[33] By the 1770s, there was no need for Newtonian mathematics to be the dominant system of mathematical expression, and Newton's calculus and Leibnizian calculus had both been superseded as systems of notations by Euler's integration.[34] Still, popular works about Newton persisted, and Newtonian contributions to science remained culturally relevant.

A book with difficult content such as the *Principia* draws attention to its own materiality because readers cannot immerse themselves in it fully, and because its response-inviting structures are at times simple to navigate, but at other times are highly specialized. If print already holds a kind of blueprint, as Louise Rosenblatt, Wolfgang Iser, and other critics have long contended, for a community of readers to interpret in varying ways, the *Principia* contains a rich template for interpretation by its readers and others whom it inspired.[35] Rosenblatt describes the reader-text relationship, in which an isolated text can be consumed, as a "transaction," a term that recalls the title of the Royal Society's periodicals, which could be consumed in one sitting, as well as the idea of a reading experience as itself a periodical one. The *Principia* could not be consumed in a single reading "transaction," and instead required a series of discrete interactions with the same text, repeated transactions that forge a relationship between reader and text over time.[36] The relationship between structure and understanding affects how scientific works are read, including the need to parse ideas before full comprehension can occur.[37] The structure, the form, the blank, and the transaction: all of these concepts help to shape the *Principia*'s reception.

Beyond thinking about the blueprint that helps shape a text's reception is the potential for readers to engage with it, which we can understand viewed through Giorgio Agamben's readings of potentiality. Agamben responds to Aristotelian potentiality, stating that it can be a kind of latency, in which the poet by definition always has the potential to write a poem.[38] This concept of latent potentiality aligns two issues that relate to this book as a whole. First, the principles of Newtonian physics I describe also theorize potentiality, so potentiality is a part of popular Newtonianism. An object's status as an object requires that it have the potential for motion. This book's focus on the materiality of books plays further with objects and their latency. A reader always has the potential to read, if not always the capacity to read and comprehend thoroughly. The information readers could potentially glean from engagement with the *Principia* is valuable to consider in its reception—not just the difficulty of the material but also the potentiality of reading. Indeed, Agamben and Aristotle find "the existence of potentiality" as a state interesting *en soi*, "that there is a presence and a face of potentiality" itself apart from looking at what that potentiality might produce.[39] The state of potentiality thus bears similarities to the history of reading, especially with reference to time and reception.

Scholars of reading privilege written evidence, in particular the written documentation of the reading experience in manuscripts, correspondence, or commonplace books, at the exclusion of other forms of evidence. I use such documented evidence of reading when it exists. I also analyze the way publishers addressed their potential readers, the books readers borrowed at libraries—and the corresponding potential for scientific autodidacticism—and the potentially far-ranging scientific information readers of poetry encountered. The works readers encounter instruct them to inhabit a mental landscape that is difficult to replicate, as research by Elaine Scarry and Andrew Piper has shown.[40] Statistical analysis like that found in William St. Clair's *The Reading Nation in the Romantic Period* is another answer to the challenges of studying the history of reading.[41] Both approaches are helpful to scholars, because the study of reading rests somewhere between the interior and the exterior: between the interiority of the reading experience and the exterior traces of readership that linger in material cultures.

The different genres of text and print contexts I am engaging with in *Reading Popular Newtonianism* require different approaches. For example, my analysis of the links between Newton's methods, authorship, the culture of late seventeenth-century science, and print in the *Principia* cannot be evaluated in the same critical framework as my analysis of the readers of popular scientific books at the New York Society Library in the late 1780s and 1790s. The two are not merely different cultures, but different cultures of print: one a place of contested intellectual priority that played out in a single book, the other a space of exchange where scientific books were borrowed and used in conjunction with printed texts in other genres by readers who sought information rather than priority.[42] Both print cultures nonetheless contribute important insights to the story of print and popular Newtonianism this book tells.[43] James Delbourgo's excellent study of early American electricity shows us how early America experimented with and responded to scientific research about electricity in ways that were both tied to cultures of British science and uniquely their own. Delbourgo shows that experimental science in early America relied heavily on Newtonian natural philosophy, especially at Harvard and in Philadelphia; it is my hope to extend this conversation by evaluating the role of print in the dissemination of Newtonian natural philosophy, using the less-studied example of New York. Eve Tavor Bannet's landmark history of transatlantic reading focuses mostly on fiction, and on other locations besides New York.[44] Susan Scott Parrish's compelling analysis of early American science focuses nearly half of its analysis on the Britishness of early American science. Parrish moves her analysis from American science's connections to the metropole, to the intersection of English bodies and American climates, to the importance of correspondence between the new and the old world.[45] My work on the New York Society Library furthers this conversation by describing the New York reception of British science, especially popular Newtonian printed works.

If we can reassemble the associative connections of a network, following Bruno Latour's work, there are tremendous implications for the history of readership, a history that has been underexplored because it is frequently undertheorized. The questions of "How do we know what people really read?" and "How do we know whether they actually understood it?" inhibit scholars from addressing readership at all. The *Principia* in particular has been underused outside of the history of science because its content is difficult. The narratives I engage with tell the story of print, readership, and popular Newtonianism across the eighteenth century. The history of reading and its utility for eighteenth-century studies rightly situates reception, as a narrative of potentiality, at the center of current scholarship.

David Brewer has shown how the reception history of fictional characters became not just about reading the original book where those characters appeared, but also about the circulation of what we would today call fan fiction and unauthorized sequels.[46] Brewer describes the long eighteenth century as "an age of perpetual negotiation between readers and writers," where the ownership of a character became a debated topic.[47] This was during the time following the Statute of Anne, when the copyright of a written text was defined clearly. Characters, Brewer shows, inhabited and penetrated British culture both from the top down and the bottom up. His analysis has in part influenced my decision to look at peripheral texts that describe Newton's ideas, including encyclopedia entries and works on aesthetics—the marginal artifacts and the central monograph help to construct one another. Newton, who may have resembled a character even more than he did a person, receives treatment similar to popular literary characters in popularizations.[48] Both Newton and literary characters circulated in printed books that aimed to enlighten diverse audiences. With the extensive reach of publications explaining scientific topics, it was impossible for the image of Newton to be controlled publicly. His character has a long print afterlife that is examined here.

The Structure of This Book

This book begins by evaluating Newton's perspective on print through my analysis of his reading, followed by my reading of print and the *Principia*. I show that even though many readers could not understand the *Principia*'s mathematics because they lacked an extensive mathematical background, the book's Latin was more accessible to its audience, and the *Principia* therefore deserves closer consideration in the narrative of Newtonian popularization. The three chapters that follow the chapter on the *Principia* focus on reading and interpretive communities—those of poetry, prose, and library borrowers.[49] The book is organized chronologically, beginning with Newton's initial publications on optics from the early 1670s, and concluding with the end of the NYSL's first charging ledger in 1792. Its narrative also progresses outward, moving from Newton's individual orientation to print to larger, decentered communities, including my analysis of an emerging public of real readers whose preferences we can trace and interpret.[50]

The first two chapters work in tandem to set up the reception of the *Principia*. Chapter 1, "The Weight of Authority in Newton's Reading and Correspondence," describes Newton's relationship to print as a reader who left traces of his own encounters with manuscripts and printed books, and introduces the print context of the *Principia* as one that links authority, community, and print. I consider ways in which other texts Newton read, wrote about, and interrogated prefigure the tensions between individuals and communities that underwrote the *Principia*'s publication history and reception. Chapter 2, "The Attractions of Print in the *Principia*," shows how Newton used print to position himself as an intellectual with a superhuman access to truth by manipulating the intelligibility of the *Principia*. From the seventeenth century until today, critics have described the *Principia* as an alienating text that can be grasped by few specialists—but writers also refer to the *Principia* as a book so important that it arguably articulates a transition to modernity.[51] I account for these contradictory attributes by arguing that even though the *Principia*'s mathematical complexity lay beyond the understanding of most readers, Newton manipulated common print conventions to make the *Principia* superficially accessible to a range of reading levels, attracting even inexpert readers to complex ideas that were presented with clarity in Latin prose.[52]

Chapter 3, "Newtonian Popularization and Masculinity in Two Poems about Newton," introduces the metaphorical hypermasculinity with which Newton was depicted in fiction, and the extent to which this representation was linked to the *Principia* and separated Newton from the contemporary scientific community. I reevaluate two poems by Edmond Halley (1656–1742) and James Thomson (1700–1748) that celebrate Newton's presence in social and scientific communities and, crucially, connect sociability and science to masculinity as embodied in print publication. Thomson, who borrowed heavily from Halley's Latin ode introducing the *Principia*, reveals the importance of Newton as a print figure as well as a scientist, noting especially the way print fashions masculinity. In the form of authorial dominance and expertise, the Newton of the *Principia* towered over an implied community of men marked inferior by their lack of expertise and access to truth.

My fourth chapter, "Extending Popularization in English-Language Editions of Francesco Algarotti's Works," reveals the diffuseness of Newtonian popularization by examining the print practices of English-language translations of Francesco Algarotti, who helped to popularize Newton. Books like Algarotti's *Il Newtonianismo per le dame* (1737; first English edition, 1739) "popularized" scientific ideas in Britain and on the Continent by translating complex ideas to a nonspecialist, literate audience. Likewise, Algarotti's essays on painting and opera helped to explain—and define—

these topics for an emerging middle class with a desire for cultural sophistication. Chapter 4 identifies the gap between Algarotti's printed works and the English-language reception of Newton, as well as the different ways Algarotti interwove scientific information with cultural and aesthetic criticism. Algarotti's influence on the reception history of science in England is, thus, more complex and related to developing print markets than has been assumed.

Chapter 5, "Reading Popular Newtonianism at the New York Society Library, 1789–1792," casts its timeline forward, exploring the development of markets for scientific reading at century's end. I analyze reading records for popular Newtonianism, using the extraordinary first charging ledger of the New York Society Library. This ledger, which consists of records for more than five hundred readers over a three-year period (1789–92), allows us to work within a developing community to reconstruct the popularity of works with Newtonian content, including popularizations and encyclopedias.

This last chapter presents a new way to understand "popularizations," contradicting the prevailing theory that people who had limited scientific instruction but an interest in science read and encountered "dumbed-down" versions of scientific works. Popular books with scientific content, popular works in general, and popularizations of science were read by the same readers and families, showing that there was substantial overlap in these categories and extensive—and surprising—dissemination of scientific information. These readers' engagements with Newtonian natural philosophy under many different banners shows that the popularization of Newton, as it diversified in the late eighteenth century, became bound to increasingly complex reading practices.

I

The Weight of Authority in Newton's Reading and Correspondence

In the *Principia* (1687), Newton synthesized print conventions and methods to position his authority such that writers cited and responded to his book in popular texts. The traces of Newton's reading in different genres link authority and print in ways that echo in his printed works, especially the *Principia*. Newton's reading records and correspondence reveal the tensions between the community and the individual that became part of the *Principia*'s publication history. The materiality of texts—central to Newton's reading—ultimately achieved a primacy that Newton would explore more fully in the *Principia*. This chapter analyzes several key moments when Newton took a position on the physicality of manuscript and print and their uses to communicate and authenticate forms of evidence. Newton's treatment of manuscript and printed evidence in the works he read and in the *Principia* evinces the limitations and the potentiality of using the printed book to convey authority.

When scholars write about a topic of interest to Newton, they tend to survey his related reading: for example, works on Newton's mathematical work engage his discussion of mathematics in correspondence and notebooks; scholars who focus on Newton's alchemy discuss his readings in alchemy. Because this book is about the dissemination of Newtonian natural philosophy—with the accompanying implication that we are looking at the public face of Newton's work across genres—I analyze Newton's reading history insofar as it relates to publicity. To this end, I focus largely on

Newton's reading in theology because his theological inquiry and his public persona intersected. Not only was Newton's work toward understanding the nature of the universe part of his theological perspective, it also inflected Newton's daily life as a professor at Trinity College, Cambridge. Newton's reading practices showed how he, like other readers of the seventeenth and early eighteenth centuries, filtered classical and contemporary sources, spoke explicitly about forms of textual evidence, and commented on the disparity between publicity and privacy—all factors that helped form the *Principia*'s publication context.

Traces of Newton's Reading

Other scholars have stressed the links between Newton's reading and his manuscripts. Jed Z. Buchwald and Mordechai Feingold delineate the process whereby Newton engaged with classical and religious history to write his *Chronology of Ancient Kingdoms.*[1] In the *Chronology*, Newton researched unusual historical texts alongside conventional classical authorities.[2] He also attempted to be systematic about what he read: whether he responded to ancient theological texts that opposed Anglican doctrine, or contemporary criticism, or the Bible itself, Newton engaged intensively with printed material and manuscripts as a reader. Ann Blair has written persuasively about how early modern readers like Newton managed the large influx of new knowledge from both contemporary and classical sources that accompanied increased print circulation.[3] Readers' notes provided valuable assistance in managing information from diverse print and manuscript sources. As in the early modern humanist reading practices Blair and Anthony Grafton have studied, readers from the late seventeenth century such as Newton chose how to filter evidence from classical sources, how to incorporate new knowledge, and how much authority to grant to ancient texts.[4] Buchwald and Feingold's study of Newton's chronology has also shown that it is useful to analyze Newton's reading in different rhetorical contexts.[5] To understand Newton's relation to the public, we will benefit from looking at his reading practices—especially those in theology—where Newton actually wrote about people's public and private identities, and the intersection of those identities with extant—and public—power structures. Understanding Newton and his public-mindedness in this capacity is essential to understanding his popularization, and contributes to our understanding of the *Principia* beyond the mathematical and scientific context that produced it.

Newton's reading across genres reflects the *Principia*'s publication context: the traces of his reading expose Newton's focus on two central topics: how to value different forms of evidence and how to evaluate publicity, especially the often tense relationship between an individual and a community.[6] These interests play out most frequently in Newton's theological reading and correspondence but are also supported by Newton's chronology work, his alchemy, and other correspondence of his that related to print publication. His focus on evidence and individual-community tensions would affect Newton's uses of mathematical evidence in the *Principia* as well as underscore the tensions between Newton the individual and the community—the Royal Society—who were both his peers and the audience that influenced his publishing decisions. The valuation of different forms of evidence and the balance between individual goals and community demands structured Newton's reading in addition to furthering his professional pursuits.[7]

The large volume of extant manuscripts related to Newton's reading makes it difficult to study his reading comprehensively.[8] None would doubt that Newton was passionate about books; he owned more than 2,100 books at the time of his death in 1727.[9] Newton had inherited more than 300 books from his stepfather; he built the rest of his collection gradually.[10] Some records of Newton's purchases and of gifts from his friends survive, but he did not make consistent notes regarding his purchase or acquisition of books. His reading is itself a magnum opus: he left behind more than 2.5 million words of manuscript writings on theology—the equivalent of 100,000 double-spaced typewritten pages today.[11] Newton's reading notes are diffusive and frequently disorganized, yet are also extensive, thoughtful, and questioning.[12] These manuscripts' sheer quantity prevented them from being used fully until the twentieth century—and even so, they compose a large textual corpus.[13]

The majority of texts Newton read were theological, mathematical, or scientific in nature, rather than literary texts, with a notable exception—Ovid's *Metamorphoses*. Newton owned Latin editions of Galileo's *Dialogue on Two Chief World Systems* and *Starry Messenger*, as well as several copies of Descartes's *Geometry*, a copy of his *Principia Philosophiae*, and a copy of his *Meditations on First Philosophy*.[14] The literary texts that he owned were older works, such as Ovid's *Metamorphoses*, and his responses to them had little to do with observing their literary merit.[15] Because of their allusions to shape-shifting and change, the *Metamorphoses* were important to practicing alchemists like Newton. Newton owned several editions of the *Metamor-*

phoses, which demonstrates his interest in the text, yet complicates analysis of how he read it. A pocket-sized Latin copy with several manuscript annotations included references to alchemical recipes, signs of heavy use, and additional marginal annotations.[16] For Newton and other alchemists, the physical transformations described in the *Metamorphoses* could serve as clues for alchemical experimentation.[17]

Because the *Metamorphoses* could be useful in alchemical practice, Newton read this work differently than a person who read for content or entertainment. For example, Newton annotates the story of Actaeon and Diana in volume 3, book 2, of the *Metamorphoses* in ways that appear to make little sense, given the story line. In the story, Actaeon sees Diana naked and she silences him, threatening to change him into a stag if he cries out. Actaeon calls for his hunting party and is changed into a stag; his dogs then attack him, killing him. This is an exciting passage; however, Newton's marginalia do not annotate the climactic moment when the dogs attack, but the section describing the various dogs that attack Actaeon. An excerpt from George Sandys's translation of the *Metamorphoses,* a copy of which Newton also owned, shows the section that was heavily annotated:

> While thus he doubts, his dogs their master view,
> *Black-foot,* and *Tracer,* opening first, pursue:
> Sure *Tracer Gnossus, Black-foot Sparta* bare.
> Then all fell in, more swift than forced Air:
> *Spie, Ravener, Clime-cliff,* these *Arcadia* bred:
> Strong *Fawn-bane, Whirl-wind,* eager *Follow-dread;*
> *Hunter,* for sent; for speed, *Flight* went before.
> Fierce *Salvage,* lately gaunched by a Boar.[18]

Rather than offer criticism on the text or editorial commentary, or even a personal reaction to the text, Newton's marginalia comprise a list in English next to the tiny volume's long Latin catalogue of the attack dogs:

> 1 blackfoot
> 2 spio.
> 3 climb-cliff
> 4 fawn-bains
> 5 whirlwind
> 6 follow-dred
> 7 flight

8 hunter
9 salvage[19]

The list continues as the passage progresses, including an additional eleven entries of dogs, including "ranger, shepherd, greedy, catch, courses, blab, patch, tiger, 'alce' strong, blanch, mourner." These footnote annotations, given in English next to a Latin text, resemble a list of vocabulary words, but are most likely an attempt to infer an alchemical recipe from the catalogue of names. Thomas Willard reinforces this interpretation: alchemists believed that "if [Ovid] knew the answers, they could be found in the poetry, no matter how much he hid them."[20] The engaging plots of the *Metamorphoses* appear to have been less noteworthy to Newton in his search for alchemical clues, and the latter parts of this particular copy of the *Metamorphoses* are unmarked.

The proliferation of copies of the *Metamorphoses* that Newton owned testifies to the hold alchemy took on his curiosity during the late seventeenth century, and exemplifies one use he made of the books he read.[21] If imaginative literature offers few strongholds for studying Newton as a leisurely reader, the imaginative use of texts and language in other forms was important to Newton's reading notes and correspondence. His awareness of material texts and their language undergirds Newton's use of print and narrative technique in the *Principia*. The traces of reading that Newton left in books show his intense physical engagement with texts: smudges, stray pen marks, odd notes, and calculations appear throughout his printed books, and Newton also dog-eared many pages for future reference.[22]

The Authority of Textual Evidence

Newton's correspondence with colleagues reveals the pressures between individuals and communities that penetrated Newton's life and exhibits the centrality of publicity to his work. Even when Newton placed himself at a remove from social circles, his letters show that he remained the subject of gossip and discussion among his peers. Rumors of his death appear in two letters from Newton's colleague and rival, Astronomer Royal John Flamsteed (1646–1719).[23] In the first letter, Flamsteed writes that "Mr. Hanway brought me news from London that you were dead but I shewed him your letter which proved the contrary. He had it from Sir C. Wren to whom he wrote immediately to satisfy him of the falsehood of that report."[24] In a later letter, Flamsteed writes that "[your letter] served me to assure your friends that you were in health they having heard that you were dead again."[25] The letters Newton wrote in response to Flamsteed become objects to be interpreted as proof that Newton himself was alive as well as proof that Flamsteed had intimate access to the celebrated, reclusive Newton. When the receipt of a letter from Newton can stand for Newton himself, this transferability indicates a slippage between persona, text, and author that I will later analyze in the reception of Newton's works.

Their correspondence took place in the years shortly after the publication of the *Principia*, so Newton's exchange with Flamsteed is contemporary to significant developments in his own public identity. If a letter could stand in for a person, a printed book like the *Principia* stands for its author and affects his reception. In spite of his colleagues' seeming interest in Newton's well-being, neither Flamsteed nor the others mentioned in his correspondence actually visited Newton to check on his health. Apart from Hanway's visit to Flamsteed where the letter was produced as evidence, the men wrote to one another in lieu of face-to-face interaction: Newton to Flamsteed, Hanway to Wren, Flamsteed to Newton. This dated manuscript letter from Newton would represent Newton, even though the actual, verifiable Newton lived a short ride away from Greenwich in London.

Beyond quieting rumors of his own death, Newton had a broader historical stake in the authenticity of letters and the use of letters as evidence for life, as his theological reading shows. Multiple manuscript copies remain of Newton's theological manuscript *"Paradoxical Questions Concerning the Morals and Actions of Athanasius and His Followers."* The origins of the Trinity in church history occupy a majority of the manuscript, which focuses on letters as proof of life and which questions textual authenticity. Robert Markley discusses this manuscript at length as "an incomplete, unabashedly heretical history" of the steps that "led to the triumph of Trinitarian doctrine in the fourth century."[26] The manuscript's theological questions regarding early Christian conflicts focus on the authenticity of a letter between Arsenius and Athanasius.[27] Athanasius and Arsenius disagreed regarding the unity of the Holy Trinity, and their correspondence would be essential to studying the Trinity's history. Markley claims that Newton's "reluctance to publish his work (noted by many of his contemporaries) may result from his frustration in never being able to make the necessary discoveries that would stabilize the relations among the semiotic systems he studies."[28] Newton analyzed conflicts in this manuscript that he also engaged with in print. Newton was indeed reluctant to publish works that were open to interpretation and criticism—works lacking in sufficient evidence—and his interest in giving the *Principia* an unassailable stability furthers this claim.

Newton's correspondence with Flamsteed seems unique—a peer astronomer expressing concern about Newton's health—yet in it Newton addresses similar concerns with style, publicity, and letter writing, seen in his reading of theological history. Newton examines the content of a letter allegedly written to Athanasius by Arsenius and determines from the letter's content and from evidence that the letter is either false or forged. He tries to place the letter's genre, which "has not the form and humour of a free letter but looks like some formal covenant of submission drawn up by a Lawyer."[29] A "free letter" is informal compared to a work with an implicit orientation to a third party, written by someone external and disinterested. Newton determines that the style of the letter is too formal to be a casual letter between acquaintances and concludes that other social forces must be behind it, such as a "magistrate" forcing a "recantation" in order to preserve Athanasius's reputation.

After he speculates that the letter is falsified because of its style, Newton extends his scrutiny to the ways in which Athanasius and the Council of Tyre evaluate the letter improperly as evidence: "they insist only upon the evidence of his letter and represent that no other evidence was to be expected and by consequence had no other, and magnify this evidence so much as if God had thereby produced Arsenius alive before the eyes of all men, and that notwithstanding this Letter the Council of Tyre proceeded against Athanasius and banished him as a murderer."[30]

Following his judgment that the letter was most likely inauthentic, Newton criticizes the misinterpretation of letters as evidence. According to this passage, Athanasius and the Bishops attempt to prove that Arsenius lives—rather than having been killed by Athanasius—based solely on the evidence of a letter. In spite of the letter, however, Athanasius is exiled for the murder of Arsenius. Newton identifies problems with both ways of treating a letter as proof of life: perhaps this is why the title "*Paradoxical Questions*" emerged after his review of the evidence. Even though a letter is not enough to acquit Athanasius, it is not enough to condemn him either. Athanasius attributes too much weight to a single letter as evidence, as the passage above indicates. A letter is not, for the reasons Newton outlines above, the same as Arius's personal testimony. The Council of Tyre responds inappropriately to this evidence by ignoring the letter and banishing Athanasius, acting as though there were no letter in the first place. Here, letters have value as evidence of a person's life, but one must be skeptical about the authority one attributes to letters.

In addition to the emphasis on textual authority and forms of evidence in the *"Paradoxical Questions,"* Newton contemplates the macabre fates of individuals who had been exiled from their communities. Newton's conflicts with the Royal Society were important in his life; these pages show that the conflicts between communities and individuals that characterized his relationship to the Society were important to his reading. Sections of this manuscript that are crossed out with a single diagonal line (here represented by a strikethrough), but which otherwise remain fully legible, show how intensely Newton pursued the story of Athanasius and Arius to its exile and death: "~~Arius being opposed by the clergy~~ . . . ~~returned to Constantinople and then died in a boghouse.~~"[31] A boghouse is a toilet or small outbuilding, where Arius, in isolation and suffering from illness, was left to slide toward death, "deprived both of communion & life. Such was the end of Arius. And the Eusebians being greatly ashamed buried their fellow conspirator, but the Church rejoycing."[32] The focus on squalor, embodiment, and death shows Newton's interest in the fates of individual transgressors at a remove from their communities.[33]

"Paradoxical Questions" also analyzes religious orthodoxy as well as the value of outward appearances to one's community. Newton contends that behaving in an orthodox fashion should mean that one is assumed to have orthodox beliefs.

> If you say they dissembled and were Arian [opposed to the Trinity] in their heart while they were orthodox in their language I must ask you how you or any man else can know that. . . . We have no other means of knowing men's faith but by profession & outward communion & way of worship: and by all the characters the fathers of this Council were orthodox. They constantly professed against Arianism and were in communion with the Churches of all the world & worshipped as other churches of that age did.[34]

Newton communicates his concern that the accused Arians of the Council of Tyre were treated as liars when they claimed to believe in the Trinity and when all their outward signs upheld this claim. This discourse on the value of personal behavior and testimony implies that Newton believed people should be judged by their actions as well as their words. This passage does not, however, refrain from skepticism about personal hypocrisy hidden under alleged community standards. Newton proceeds to compare the Council of Tyre and the present, making claims about seventeenth-century morés. He writes that if his own contemporaries were accused of similar crimes—even if they appeared to be upright citizens who were not decried previously by their enemies—"I think such proceedings would by all sober men be accounted as malicious & barbarous as any we ever heard of."[35] In this capacity, Newton's present life informed his reading of past manuscripts.

In spite of his social reticence and resistance to community pressures, Newton's correspondence related to theology presented an epistolary form of sociability, and as such, we can understand his perspectives on publicity from examining it. Newton's relationship with John Locke pivoted on interrogating complex philosophical and theological topics through correspondence: Locke is also one of the central figures discussed in Richard Yeo's recent book on science and the notebooks of virtuosi, so manuscript accounts were important to the philosopher as well.[36] Newton corresponded with Locke regarding one of the epistles of St. Paul to the Corinthians in which both Locke and Newton had been confused by terminology. The verse under discussion, I Corinthians 7:14, states, "For the unbelieving husband is sanctified by the wife, and the unbelieving wife is sanctified by the husband: else were your children unclean, but now are they holy." The source of confusion for Newton and Locke was that this verse contradicts other Bible verses because it implies redemption through marriage, rather than through a conscious act of conversion. Taken at face value, the verse exposes a contradiction in divine laws.

The letter to Locke identifies what Newton sees as the central contradiction behind this Bible verse—one that relates to definitions of community. Newton attends to the motives underlying the terms of "unbelieving" and "sanctified" within the verse, interpreting "unbelieving" and "sanctified" not as labels that would mark people as converted by marriage, but as necessary exemptions to divine law to ensure that Christianity would spread over generations as a community-building force. Newton writes that "the believing wife must not separate from the unbelieving husband as unholy or unclean nor the believing husband from the nonbelieving wife, for the unbeliever is sanctified or cleansed by marriage with the believer, the law of avoiding the company of unbelievers being in this case dispensed with."[37] In this logical conundrum, people who are unclean must not be associated with, but then religion cannot spread if all potential spouses must be avoided as unclean. As with Newton's previous interest in the community standards for behavior demonstrated in the Council of Tyre, his reading of this paradoxical question from I Corinthians contemplates the ways communities dictate the standards for individual behavior.

In addition to his correspondence with Locke, Newton's annotations to his Bible reflected an interest in individuals' relationship to community. Newton annotated pages of his Bible throughout his life, at times leaving multiple cross-references next to a single verse, while other times going several pages without a mark.[38] His marginalia to Deuteronomy 20–22 offer typical examples of Newton's biblical annotations. The passage contains some markings, but not as extensive marginalia as some of his other Bible verses. Even so, this typical page weighs in on evidence, community standards, and individual pressures—also in a discussion of marriage. The only annotation on this page accompanies the section of the text that refers to paternal responsibilities in the event that one marries one's daughter to a sexually unhappy spouse: "The damsel's father shall say unto the elders, I gave my daughter unto this man to wife, and he hateth her; And, lo, he hath given occasions of speech against her, saying, I found not thy daughter a maid; and yet these are the tokens of my daughter's virginity. And they shall spread the cloth before the elders of the city. And the elders of that city shall take that man and chastise him."[39] Like many passages of the Bible, these lines offer guidelines for community behavior in the event of an individual's sin. As in the I Corinthians passage described earlier, where marriage partners are categorized as unclean or clean, this passage illustrates the harsh treatment of individuals who threaten the community. In the prior case, marriage with unclean partners was sanctified, but in this passage, the individual who threatens the community by lying about consummating his marriage is chastised for abandoning his wife and is punished by the community. Newton focuses on the severe punishment to be dealt to individuals who threaten community order as he had done in the passage on Arius. Here, the weight of authority also turns on evidence found on a sheet—in this instance, a cloth that reveals sexual consummation—to chastise the man who threatens his community. The man who has "given occasions of speech" cannot out-argue the "tokens" of physical evidence a community values.

Newton's manuscript annotation to these lines directs the reader's attention to Deuteronomy 16:16, which does not seem to have an immediate application to Deuteronomy 22 because it is about feasts; however, the passage also focuses on men who police their communities, stating that "three times in a year shall all thy males appear before the Lord thy God in the place which he shall choose; in the feast of unleavened bread, and in the feast of weeks, and in the feast of tabernacles: and they shall not appear before the Lord empty."[40] These lines describe feasts that occur three times per year, where all males gather together in communal ritual. Even though these two passages seem disparate in tone and content, their centrality to Newton's reading makes sense. Together, the two passages evoke a world where men collaborate to maintain the boundaries of their community and serve God. In the first instance, the community serves God by appropriating physical evidence correctly. The seducer of a man's daughter has ignored the physical evidence that he has seduced her and instead undermined her reputation with unsubstantiated testimony; it is the responsibility of the community, in this instance, to bring justice to the father and to his wronged daughter. The second passage shows that it is the community's responsibility to serve and honor God.

Communities and their formation were relevant topics for Newton, who set himself apart from his peers in publication, and who also lived in a homosocial community during his early career as an academic at Cambridge. Like the earlier priests Newton read about, academics were a community set apart, much as, Newton writes in another manuscript, "priests anciently were above other men well skilled in the knowledge of the true frame of Nature."[41] Other communities of which Newton was a part experienced conflict, in his early years as a Fellow of the Royal Society, and in later years when Newton served as the Society's president. Buchwald and Feingold describe Newton's relationship to religious community as one that was connected to his reading in multiple intellectual fields: both Isaac Barrow and John Locke had spoken out about the tension between pursuing the religious life expected of a scholar and the desire to read and research more widely. For Newton to take holy orders—the normal path of someone at Cambridge—diverged from the motivation to pursue mathematical and scientific interests. This would not have been a trivial decision, in which the duties of the clergy would have been an annoyance or a distraction—rather, Barrow, Newton, and Locke took religion seriously, and knew that they could not pursue a career in the clergy without full commitment.[42]

Newton's breadth as a thinker and reader, as well as his suspicion of print publication, surfaced in his work on chronology. In order to write a chronology, one must synthesize dissimilar texts and unite them systemically, so Newton's *Chronology of Ancient Kingdoms Amended* was composed of history, theology, and mathematics. Another member of Newton's community, Locke, likewise stressed the importance of chronology in education. Locke's attention to chronology supported his own interest in broad, well-rounded education, so chronology projects like those on which Newton worked align with a Lockean model of education. In the 1720s, Newton intervened to prevent the unauthorized publication of an early draft of his *Chronology of Ancient Kingdoms Amended.* This historical chronology reframed biblical history according to Newton's private calculations of when different events occurred, exhibiting Newton's diverse interests in the history of religion as well as his mathematical skills.[43] The *Chronology* began as a manuscript, not a personal letter, but because of Newton's investment in his own reputation, it was protected as closely as if it were a personal confession. Even near the end of his life—he died in 1727—many of the concerns about publicity and community pressures that occupied his philosophical, personal, and religious correspondence still resonated in Newton's writing, especially when it came to unauthorized printing of his works.

A manuscript of Newton's related to the unauthorized *Chronology* publication consists of no fewer than seven drafts of Newton's "Defence of the *Chronology of Ancient Kingdoms.*" His desire to respond publicly to the unauthorized publication of this text was so strong that Newton drafted a response to the unauthorized French publisher seven times. One such draft barely conceals Newton's scorn at the suggestion that he would consent to the publication. He writes that "the Antiquary has got a Printer to print the translation & the confutation, & the Printer has endeavoured to get my leave to print the translation without sending me a copy thereof to peruse it, or letting me know that his designe was to print \it/ with a confutation unanswered & unknown to me."[44] Here Newton is especially agitated that he will not have a proof of the manuscript to examine—and that there will be a confutation added to it that Newton also does not know. As with Newton's detective work on Arsenius and Athanasius, where a physical copy of a letter could represent a person, Newton wants proofs to review, so that he will understand what has been pirated, and what—if anything—he might choose to do in response. In my other examples, the presence of material evidence had offered positive proof—of consummation, of Newton's vitality, of someone's life. In this instance, the absence of textual evidence deflects control from Newton, replacing it with helpless frustration.

In Newton's reading, texts served diverse, complex functions, allowing

Newton to examine the valuation of evidence, the standards that bound communities together, and the primacy of the material text. Newton's private reading was connected to public discourse, and reveals the importance of privacy to an individual's intellectual pursuits. Spacks writes that "eighteenth-century England did not originate the concept of privacy, but the evidence indicates a new level of attention to it during the period."[45] This heightened focus on privacy manifests in reading notes and correspondence like Newton's, where a person's interior encounter with reading a text becomes an external object, a manuscript. Francis Barker describes Pepys's *Diary* as "for all the fullness of its days, despite being so richly populated with others and with the furniture of gossip and events, [ ] the record of a terrible isolation."[46] Unlike Pepys, who engaged in an active social life that suffused his *Diary*, Newton was inclined to privacy, often refusing to travel even to London from Cambridge for meetings of the Royal Society, and, as Flamsteed's correspondence shows, he could seem to lead the life of a hermit even in London. In light of these examples from Newton's theological reading, there was nothing terrible or necessarily isolating about Newton's extensive reading records: they were linked to his perspectives on publicity and community. These perspectives facilitated the *Principia*'s publication.

The *Principia* in Print History

Now that I have examined the centrality of debates regarding publicity in Newton's reading records, I turn to the *Principia*'s context adjacent to the *Philosophical Transactions of the Royal Society* in order to better understand how the *Principia* connects to the other printed works Newton and his contemporaries read. The Society's motto, *Nullius in Verba* ("Nothing in Words"), implies that truth lies in what can be demonstrated, paradoxically using words to criticize language's ability to communicate truth. Words were necessary to the Royal Society.[47] In order to publish his early papers on optics in the *Philosophical Transactions*, Newton modeled his writing on the rhetorical style the Society favored, where the description of an experiment facilitated virtual witnessing. As would later be apparent in the *Principia*, Newton crafted his essays for the *Transactions* with great care. What this meant in periodical publication was that Newton's attempts to mimic the *Transactions*' preferred style meant minimizing extensive experimentation and a broad research agenda as if it were a single day's experiment: "In the beginning of the year 1666 . . . I procured me a triangular glass prism, to try therewith the celebrated phaenomena of colours. And in order thereto having darkened my chamber, and made a small hole in my window-shuts, to let in a convenient quantity of the sun's light, I placed my prism at his entrance, that it might be thereby refracted to the opposite wall. It was at first a very pleasing divertisement, to view the vivid and intense colors produced thereby."[48] Newton describes an optical experiment where he manipulates the angles of two prisms placed in a darkened room with a narrow entrance of light, causing the light to separate out into the spectrum on the wall.

Although Newton's prose follows the first-person narrative favored by the Royal Society, for him to do so meant exaggerating parts of the experiment's research timeline. Newton was not simply "trying phaenomena" one day; the phenomena Newton observed purported to advance current optical theory regarding the properties of light.[49] Rather than sitting in front of the spectrum, "applying myself to consider [the colors] more circumspectly," Newton performed numerous experiments in order to achieve the correct angle of the prism and the perfect amount of daylight.[50] The prose style Newton used devalued the experiment's significance, labeling its results "a very pleasing divertisement," as if knowledge and pleasure were interchangeable. The temporality of the story published in the *Transactions* is fleeting, the labor of a sunny afternoon, rather than a series of experiments connected to a broader research agenda that were conducted over a long period of time.[51] Newton would work to counterbalance this intersection of "divertisement" and labor with the exposition of mathematical and experimental labor in the *Principia*.

Newton chose Latin, the *lingua franca* of the European intellectual community, as the language of the *Principia*, and his reading notes and notebooks were written in both Latin and English. As early as the 1660s, as his Trinity College notebooks show, Newton was categorizing information in ways similar to his treatment of physics in the *Principia*. His notebook entry dated May 16, 1666, stated that "to resolve Problems by motion the 6 following propositions are necessary & sufficient," which proceeds to an early description of the method of fluxions, Newton's term for derivatives.[52] He would later write in Latin about the importance of this manuscript, and its significance to the theory of fluxions: "In alio Manuscripto 16 Maij 1666 composito, mathodum solvendi problemata per motum, complexus sum Propositionibus septem, quarum ultima est Regula jam descripta eliciendi velocit[at]es crescendo vel decrescendo ex aequatione quantitates crescents vel decrescentes involvente."[53] Newton's ability to move back and forth between Latin and English composition came from his early education and his later frequent use of Latin at Cambridge. Westfall rightly credits Newton's foundation at Grantham's grammar school with his proficiency in Latin, observing that because Newton's readings in natural philosophy at Cambridge were often in Latin, he was able to "communicate [easily] with European science."[54]

In some manuscripts, Newton switched back and forth between English and Latin, including his "Theologiae Gentilis Origins Philosophicae" (Philosophical origins of Gentile theology), which his biographer Westfall describes as "chaotic manuscript remains."[55] This manuscript even contained a page headed by a tangle of Latin and English: "De motu Fluid," which "suggests a rejected sheet from the *Principia*," rather than a wholly theological manuscript.[56] Newton's manuscripts as a corpus alternate between theology, natural philosophy, and alchemy, with each of these three occupying his attention at different stages of his career.

Elsewhere, in a draft of "An Account of the Systeme of the World Described in Mr. Newton's Mathematicall Principles of Philosophy," Newton wrote a very simplified response to those who used either scripture or mathematics as proof of the geocentric model. Newton's short corrective engaged multiple topics, including theology, mathematics, and mechanics. Its three headings were as follows: "I Scripture abused to prove the immoveableness of the globe of the Earth. II Mathematicks abused to prove the Globe of the Earth immoveable, [and] III Accurate skill in geometry & Mechanicks requisite to decide the Question."[57] What is important to note is that this account is aimed at public perceptions based on religion; it is organized in three books, as the *Principia* had been; and, in three short paragraphs that use clear language, it explains some of Newton's principles. Because it is an unpublished manuscript, we cannot analyze the impact this explanation had on Newtonian popularization, but it is evidence that Newton considered the popular impact of the *Principia*. By drafting a response to his own text that engaged what we would today call multiple disciplines, Newton anticipated negative reactions from readers who would need simple explanations, the kinds of readers who would desire the popularizations I examine later in this book.

Some of Newton's correspondence related to printed works he himself had published, and reflects the focus on authority and evidence exemplified in his manuscripts. Newton's early optical publications in the *Transactions* shaped his ideas about publishing future works when his writing provoked a hasty and dismissive response by the Royal Society's curator of experiments, Robert Hooke.[58] Hooke's criticism made an easy enemy of the irritable Newton.[59] The angered Newton took months to reply so that he could include necessary support for his claims. Newton even attacked the established Hooke's own work in his reply, and did not publish his own *Opticks* (1704) as a monograph until after Hooke's death, when Newton was president of the Royal Society.[60] Newton's later decision to publish the *Principia* as a monograph stemmed from the idea that shorter-form academic writing could cede the control of the reception of intensely considered, professional projects like his to "smatterers . . . those that will call themselves philosophers without mathematics."[61] Even as it was a popular and useful mode of disseminating information, the *Philosophical Transactions* privileged concision and did not allow for the front matter or other paratexts of a monograph. In spite of these limitations for writers like Newton, the *Transactions* were more fully accessible to a larger number of people than the *Principia* was, so the decision to publish outside of the *Transactions* represented a choice to expand the textual reach of Newtonian ideas while controlling rather than eliminating their popular reach.[62] Scholars describe the Royal Society's privileging of outward appearances, such as demonstration and experiment—and even sociability—as connected to truth.[63] Newton embraced experiment, but he also stressed the importance of providing a systemic foundation for experiment, bridging philosophy to the Society's ongoing project of natural inquiry, because certainty and first principles had not always accompanied witnessed experiment and the manufacture of scientific instruments.[64]

The contents of the *Philosophical Transactions* in 1685, while Newton was working on the *Principia*, display a diverse body of knowledge among its contributors.[65] As a Fellow of the Royal Society who kept abreast of current work, Newton would have been aware of what was in the *Transactions*. Even a single issue demonstrates heterogeneity in research. The first listing from this volume, "Some Experiments about Freezing, and the Difference betwixt Fresh-water Ice, and That of Sea-water; also a Probable Conjecture about the Original of the *Nitre* of Aegypt: By Dr. Lister," offers two different subjects within the same article—the properties of frozen water and an archaeological "probable conjecture."[66] Other entries from this volume include a "New Way of Cleaving Rocks," an account of a new substance "found between the pericardium and heart of an Ox," and an experiment regarding the proper size of coach wheels for being "drawn over stones etc. lying in the way."[67] These represent some competing authorial voices from which Newton would have had to distinguish himself had he published his research on Kepler's inverse square law and universal gravitation in the *Transactions* instead of as the monograph *Principia*. The more speedily scientific reputations could be made through periodicals, the more problems emerged for natural philosophers like Newton who wished to develop full mathematical proofs including multiple scholia and lemmata before submitting their research to public scrutiny. The *Transactions*' smaller, chattier format opened up intensely considered projects to the "smatterers" Newton disliked.[68]

The *Philosophical Transactions*' print history makes the study of its reception a complex subject. The formal arrangement of articles, in which diverse experiments are placed next to one another without rank, diffuses the impact that significant articles would appear to have. In addition, the medium of print itself flattens out differences in the perception of individual experiments made across time. Lastly, print elides differences in instrumental expertise among experimenters (including the differences in the manufacture or scale of lenses and prisms) and differences in accuracy among experiments.[69] In the *Transactions*, principles of comparison and evaluative rubrics were implicitly the province of its readers, whose expertise and interest varied. The reading process involved alternating immersion, erasure, and reimmersion, with the readers of the *Transactions* serving as their own textual intermediaries. Because at least part of this readership became the readership of the *Principia*, we must remember how well the *Transactions* exemplify the degree to which expertise and enthusiasm worked together in late seventeenth-century British scientific publishing.[70]

Beyond their eclectic tables of contents, the *Transactions* were under-vetted and subject to piracy. There were texts published on scientific matters with identical paginations, size, and signatures as the *Philosophical Transactions*.[71] The visual presentation of print, then, was not enough to convey authority without content that was worthy of it. Piracy could devalue the credibility of the publication, another factor we have seen was very important to Newton. Its contributors also used publication in the *Transactions* to claim public priority: this, too, would prove to be problematic, given the *Transactions*' formal leveling of epistemological value.

To select a printer for the *Principia*, Halley and Newton chose Joseph Streater, who had worked on other projects for the Royal Society, including the *Philosophical Transactions*. Streater printed other works that were affiliated with the Royal Society, including Perrault's *Memoirs for a Natural History of Animals* (1687–88), an illustrated description of the exotic animals in Louis XIV's menagerie.[72] These works' materiality gestures to their audience, much like the *Principia* would do. The dedication page of Perrault is similar to the dedication page of Newton's *Principia*, which is not surprising, considering that Streater printed both works. The translation of Perrault, however, contains a lengthier and more precise endorsement by the Royal Society. The imprimatur is placed on a separate page from the title page: "November the 17th, 1686. At a council of the Royal Society this Day held, it is ordered that the Translation of the *Memoires pour servir a l'histoire des animaux* (By Alexander Pitfeild, Esq., R.S.S.) be allowed to be printed. Tho. Henshaw, Vic. Praes. S. R." This edition was contemporary to the *Principia*'s printing in 1687, and the same person printed it, yet its imprimatur is more explicit and elaborate than the *Principia*'s.[73] The *Principia* merely lists the Royal Society's then acting president Samuel Pepys on the title page, whereas the translation of Perrault's text "is ordered" to be printed. This association of the Royal Society's imprimatur with elite status closely approximates the role of the Society's imprimatur in the *Principia*. Streater's office was overwhelmed by the printing of the *Principia:* the "Advertisement" following Halley's review of the *Principia* in the *Philosophical Transactions* apologizes for his delay in printing the *Transactions*. In fact, the *Transactions* itself states that printing the *Principia* "hath been more serviceable to the Commonwealth of Learning" than printing the *Transactions*.[74]

If we are to look solely at its printer, the *Principia*'s academic associations lower somewhat, because Streater also printed pornography; his company was prosecuted as "the capital's leading producer of lewd books."[75] Streater's pornography illuminates further the publication climate the *Principia* entered. Streater later printed the Earl of Rochester's *The Farce of Sodom, or The Quintessence of Debauchery* (1689). Like the *Principia*, *The Farce of Sodom* was controversial; unlike the *Principia*, *The Farce of Sodom* had a scandalous effect on its publisher, Benjamin Crayle, who worked with Streater. The book was ordered banned and destroyed, and the fine of twenty pounds was debilitating to Crayle's business. The text shares some thematic content with the *Principia*. Both works investigate whether it is better for one person to innovate or for a community to do so. The play contains a "Virtuoso" character through whom experimental culture is criticized. Like Thomas Shadwell's eponymous character from *The Virtuoso* (1676), Rochester's virtuoso is mocked for his lack of sexual knowledge. Not surprisingly, the play supports experimentation and reliance on the organs of sense perception in combination with intellect, with the proviso that research agendas be libertine in their origins.[76] In 1687, Streater was disgraced publicly for printing pornography: this scandal was enough that the *Principia* was eventually reprinted without Streater's name on the title page.[77] However, although Rochester's book was pornographic, its author was a nobleman, and pornography—especially pornographic satire—was a product for gentlemen with disposable income. We should not assume that the *Principia* was from a seedy printer as much as from a printer who addressed readers of similar social classes at different levels of publicity. Streater's smaller house was a good fit for a smaller book that had a specialized audience like the *Principia*, which was not expected to sell many copies, and which had an initial print run of only 300–400 copies.[78] This small print run indicates that the dissemination of Newtonian natural philosophy in large numbers was not assumed at the book's initial printing. However, that number of copies could make a great difference if placed in the right hands, which is exactly what happened.

Before I proceed to a more detailed examination of it in the next chapter, I define the editorial history of the *Principia* as having four stages. Its multi-stage editorial history reflects Newton's dedication to self-presentation: many of each edition's changes and inclusions concerned its accessibility to readers. The first stage dates from before its initial publication in 1687, when manuscripts circulated between Halley and Newton. The second stage dates from the publication of the first edition in 1687 to the publication of the second edition in 1713. The third stage dates from the publication of the second edition to the publication of the third edition in 1726, the last Latin edition Newton authorized. Each edition Newton authorized contained substantial changes, and the timing of the editions suggests that the Statute of Anne—in which a post-1710 copyright would last for fourteen years—played a part in Newton's choice to publish his last two editions thirteen years apart.[79] The last stage of publication for the *Principia* begins after Newton's death; this stage includes the publication of its first English edition, translated by Andrew Motte, which was published by his brother Benjamin in 1729, two years following Newton's death. This stage marks the end of the Latin-only *Principia*, and the end of Newton's editorial control. My book focuses principally on the Latin editions that were published during Newton's lifetime because these were the editions Newton himself oversaw and therefore were most closely connected to his self-presentation through print.

From the *Principia*'s first edition on, Newton paid special attention to the details of the printed text. Extant correspondence with Halley mentions this attention to minutiae in Newton's corrected proofs to the first edition. Newton wrote to Halley that he wanted to take his time with the editorial and proofreading process: "The printed sheets I intend to looke over. . . . I wish the printer be carefull to mend all you note. Sir I am very sensible of the great trouble you are at in this business, & the great care you take about it. Pray take your own time. And if you meet with any thing else which you think need either correcting or further explaining be pleased to signify it to Your humble & Obliged Servant Is. Newton."[80] Newton writes that he does not mind slowing down the publication process if it means the book will be published to his satisfaction, and even at the proof stage implies that he remains open to new feedback from Halley. This was no small feat during the hand-press era, when a work like the *Principia* required careful typesetting due to its complex layout and its interlaid illustrations. Newton's desire for the cohesion of a monograph would outlive him when his *Prophecies* were published in the 1730s, the editor writing that "as it is to [be] printed it will surely give more satisfaction and better carry on the thread to have everything in its own place than broken to pieces."[81] Using language echoing the equilibrium described in the *Principia*, where physical cohesion implies even, and therefore desirable, gravitational forces, these lines describe narrative preference and concern for a work's reception as well as Newton's ideas about what should be put into print.

Newton manipulated the image the public associated with his writing because it was essential to his career, both early in life as a professor at Cambridge, and later, as he took on the public roles of president of the Royal Society and Master of the Mint. Robert Iliffe describes some of the contrast between Newton's private identity and his public image as an "idol," a kind of idolatry that we will see was facilitated by the tensions between print dissemination and community. Manushag N. Powell's book on the "eidolons" of periodical literature—the authorial personae of periodical writers—also shows how individuals used print to fashion themselves in a public image for a wider community.[82] Powell writes that "to entice their audiences to read, and thereby ensure both their own paychecks and the continued survival of the medium, authors began to offer up more than advice: they offered up themselves, or rather, they offered up a version of 'the author' to be taken and mistaken for themselves."[83] Powell's comments apply to both periodical writers and scientific writers, although the scientific writers were less interested in "enticing" than in persuading audiences. Newton's self-presentation as an authoritative mathematician and philosopher in the *Principia* is one that readers both took at face value and misused to apply to his personal life. This misidentification of "real" Newton with "author" Newton accounts for representations of Newton that diverge from what we know of his biography. Newton—the reclusive, never-married, contentious scholar—was portrayed, variously, as a virile seducer of Mother Nature and a demigod. This, perhaps, was the price of privacy for Newton—a glorified public identity that has failed to align with what we know of his private life.[84] The next chapter will describe Newton's deft manipulation of the print medium in the *Principia* to cement his authority in the scientific community and guide his reception in printed works. Newton created a tiered accessibility that provided some transparency for readers who had varying mathematical expertise, without ceding his own authority or losing control of his work's dissemination.

2 The Attractions of Print in the *Principia*

Chapter 1 explored how Newton wrote about publicity through his recorded encounters with printed works. In keeping with this book's focus on popular Newtonianism and reading, this chapter shows how Newton's print strategies helped orient his work toward the popular. Newton used print practices to construct tiered accessibility for the *Principia*'s readers. To do so, he manipulated the *Principia*'s language, images, typography, form, and paratexts. These deeper structures of the *Principia* facilitated Newtonian popularization because readers could use their familiarity with print conventions to make sense of a challenging Latin mathematical treatise.[1] The *Principia*'s narrative represented the culmination of Newton's perspectives toward a print community and was essential to Newton's print reception across popularizations and genres. Since the time of its publication, critics have referred to the *Principia* as "incomprehensible" and "unintelligible"; it has been considered a text that can be grasped only by specialists.[2] This would be the case if we were to look at the *Principia* as a text that must be understood fully or not at all—and this might have been the case for some eighteenth-century practitioners of Newtonian natural philosophy. However, recent critics have taken a more nuanced approach to evaluating the intelligibility of the *Principia,* as I intend to do here. For example, Robert Iliffe describes the immediate reception of the *Principia* as structured by different kinds of understanding: the foremost questions were who understood the book, who didn't, and to what degree they understood Newton's text. Iliffe writes that "such processes constituted a sort of concentric ring of competence extending inward from those who were held to understand nothing, to the center point inhabited—perhaps—by the author himself."[3] Even as the *Principia* was a difficult book in the center of those concentric rings, it was also able to be read superficially by readers without a lot of mathematical background who had some grounding in Latin. Newton enabled this comprehension through his strategic use of print.

The *Principia* contained a mathematical explanation of the universe, including the mathematical union of celestial and terrestrial phenomena and a corresponding proof of universal gravitation. Newton demonstrated that the forces governing both celestial gravity and terrestrial gravity were mathematically equivalent and thus universal. Newton constructed the *Principia*'s front matter, its consistent narrative, and its images to be formally cohesive. The phenomena Iliffe describes were, I argue, largely facilitated by this deft manipulation of the print medium. By "intelligibility," borrowing Iliffe's term, I refer to the fact that much could be gleaned from a reading of the *Principia* even given a reader's limited knowledge of Latin and mathematics.[4] The medium of print assimilated methods like mathematical extrapolation as well as less conventional perturbation problems into a single, uniform material object—the book. The visuality and abstraction made possible through the print monograph helped to integrate Newton into popular culture and literature during the eighteenth century, as subsequent chapters show.[5]

Even though many readers could not understand the *Principia*'s mathematics because they lacked a mathematical background, the book's Latin was more comprehensible to them. We can use a minimum baseline for understanding the *Principia*'s composite reader. Anyone who tried to read Newton's book would have had some literacy in Latin. Many, but not all, would have had at least a rudimentary mathematical education. All would understand how to navigate a printed text. By granting these minimal parameters to his readership rather than comparing the knowledge of the book's readers to the almost unparalleled intellect of its writer, the extent to which readers understood the *Principia* becomes something to interpret rather than reject. As Penelope Wilson points out in her analysis of classical poetry's eighteenth-century readers, readers of classical texts were a "restricted group . . . which for the most part received a strongly classical education under private tutors, at school, and sometimes at university."[6] These readers engaged with other Latin texts, including the *Principia*. This group of readers was restricted, but these readers were an important component of Newtonian popularization because they helped to reify Newton's authority.

The *Principia*'s print context contrasted to that of the *Philosophical Transactions of the Royal Society*, which privileged prose that replicated empirical perception such that a reader became a virtual witness. The *Principia*'s print and language made abstract entities concrete and unified, as part of what Newton would posit as the universe's consistent design. The *Principia* did not substitute vivid language for vividity; instead, it used language to embody the invisible and the visible together. The idea of the "modest witness," famously put forth by Steven Shapin and Simon Schaffer and further refined by Donna Haraway and Tita Chico, works differently when reading the *Principia* than it does when reading—or "witnessing virtually"—the *Philosophical Transactions*. Haraway writes that

> the modest witness is the legitimate and authorized ventriloquist for the object world, adding nothing from his mere opinions, from his biasing embodiment. And so he is endowed with the remarkable power to establish the facts. He bears witness: he is objective; he guarantees the clarity and purity of objects. His subjectivity is his objectivity. His narratives have a magical power—they lose all trace of their history as stories, as products of partisan projects, as contestable representations, or as constructed documents in their potent capacity to define the facts.[7]

The *Principia* cannot qualify Newton as a modest witness, according to these terms, even if some of them do apply to his work. Newton is modest in the introduction to the *Principia* insofar as he does not champion his own accomplishments; the phrase "endowed with the remarkable power to establish the facts" applies both to Newton and to Edmond Halley. Halley pushed Newton to publish the *Principia* and wrote a prefatory Latin ode celebrating Newton and investing him with that level of authority. Even Halley, however, is far from modest in his prefatory ode on Newton. Instead, the principal witnessing described in the ode is Newton's own witnessing to the realm of the gods, a kind of witnessing that is at once immodest—it includes the seduction of feminized celestial entities—and aspirational rather than humble.[8]

The most notable exception to modest witnessing in the *Principia* relates to print and narrative. Consider Haraway's words "constructed documents." If a text must seem to lack narrative construction for a witness to be modest—the Royal Society's motto, *Nullius in Verba* ("Nothing in Words"), reflecting freedom from artifice—then a rigidly organized volume like the *Principia* fails this test, because its penetrability requires that readers apply the rules of its form as well as its content in order to understand it.[9] Haraway's analysis in the above-quoted lines of the material, social, and literary technologies employed by Robert Boyle resonates with the functioning of technologies in the *Principia*. The material technology of the *Principia* is the printed book. The social technology of the *Principia* includes print circulation and social exchanges oriented around a printed work, such as correspondence and discussion. The literary technology of the *Principia* incorporates a plain style of writing as the *Transactions* did, but in Latin rather than in English. This plainness extends to its use of Euclidean geometry as standard practice, rather than calculus or Cartesian geometry. Much as Haraway describes Boyle's work and the work of the late seventeenth-century Royal Society as literary innovation, Newton's writing was likewise constructed deliberately.

Although the *Principia* is one of the most challenging texts of early modern science, many of its concepts are not complex. But to contemplate their aggregate is challenging—the way that Newton built from the three laws of motion to his proof of universal gravitation meant that problems became increasingly difficult as readers made their way through Newton's book.[10] John Bender and Robin Valenza address this difficulty when they write about the *Principia,* including Bender's chapter on hypotheses from *Ends of Enlightenment* and Valenza's chapter on Newton in *Literature, Language, and the Rise of the Intellectual Disciplines.*[11] Certain principles transcended the difficulty of the *Principia*'s content and helped secure its reputation among readers, including its theorization of extrapolation, the concept of action at a distance, the placement of Newton as separated from communities of men and linked to divine authority, and the attention to form and style. The text benefited from print practices such as paratexts, editing, typography, and visuals, all of which enhanced the presentation of these principles and facilitated Newtonian popularization.

Newton was highly attentive to the *Principia*'s potential reception before it was even published; its structure and organization evolved significantly as he was writing it. It was initially entitled *"De Motu Corporum"* (*"On the Motion of Bodies"*), with two sections: "Lectures on Motion," and "The System of the World." The informal, personal-sounding title "Lectures on Motion" would disappear when the text was expanded before publication into three sections. The first book, "De Motu Corporum," focuses on the motion of bodies in unresisting space. In it, Newton established the mathematical properties of objects in motion without having to include an external force (such as gravity or water) affecting the motion of these objects. The second book builds on the proofs of the first book in order to theorize the motion of bodies in resistant media, such as water, and also included Newton's theory of hydraulics. Book 2 aimed to disprove Descartes's theory of vortices from the *Principia Philosophiae* and therefore required more space and attention than would be assumed from the subject matter alone.[12] The third book, entitled "De Mundi Systemate" ("The System of the World"), builds upon the mathematical formulae of the first two, calculating motion, not just within a resistant medium, but within the context of universal gravitation, applying the mathematical formulae from the first two books to the movements of planets in the solar system, as well as the moon's orbit around the earth. "The System of the World" was originally meant to be much easier to understand, but Newton rewrote it to make sure that only those who understood the first two books could understand it fully. The third book was the most controversial, in that it proved universal gravitation and used the same mathematical formulae to bridge phenomena on earth to those in the heavens.[13] Because of these innovations, the third book of the *Principia* was the most important book of which to claim authorship.

In 1713 Newton added a section following book 3, a "General Scholium" to the *Principia*'s second edition. The General Scholium concludes the *Principia* with a nonmathematical narrative of how the rules Newton has proven connect to God's creation and divine authority. This addition therefore increases the *Principia*'s audience engagement. The book's tightly constructed narrative, where mathematical complexity builds as the factors that inhibit motion build, traces the focus of Newton's formulae as they shift outward from small objects to the universe. In order to fully impress the significance of book 3, Newton had to emphasize the transparent qualities of his front matter. The first pages of the *Principia* bore a strong resemblance to Latin that students would have learned—especially Halley's opening ode. Those with knowledge of Euclidean geometry had further advantage in understanding the *Principia*.[14] In addition to his careful choice of mathematical and editorial practices, Newton's use of extrapolation plays a critical role in the work's narrative. The method of extrapolation—where the laws and formulae that apply to experimental evidence also apply to celestial phenomena—made the *Principia* more accessible to its audience because extrapolation was applied consistently as a method throughout the text. The narrative structure of the *Principia* mirrors Newton's method of extrapolation: it moves from small, one-body problems to multi-body problems incrementally, and moves to grander, celestial problems in much the same way.

In the section of definitions that begins the *Principia*, Newton writes that "although time, space, place and motion are very familiar to everyone, it must be noted that each of these qualities are popularly conceived solely with reference to the objects of sense perception. And this is the source of certain preconceptions."[15] According to Newton, the objects of sense perception, including the phenomena people witness, are not absolute truths and have, in fact, misled people from those truths. The experiments researchers conduct may obey universal laws, but the truths one derives from them must also be mathematically comparable. Understanding the fact that time and space and motion exist and are the same *everywhere* brings readers a step closer to understanding that time, space, and motion on earth and in space correspond. The fact that the same word, "space," is used today to refer to both celestial and terrestrial space shows how much the type of extrapolation Newton used has become normalized.[16] From these definitions of absolutes, the *Principia* proceeds to the main subject of its study, "a variety of forces and the different kinds of motions they produce."[17] Newton proceeds to describe the axioms, his three laws of motion. As with the definitions of absolutes, the three laws of motion are relatively simple to comprehend. Even within the three laws of motion, he proceeds from small to large, from simple to complex, reflecting mathematical extrapolation in narrative. His discussion of time and absolutes in the scholium to the definitions also moves from simple to complex: "Hora, Dies, Mensis, Annus."[18]

Within this formal structure, Newton developed a style of proof that offered unique resolution to a conflict between competing forms of evidence. The forms of proof that were acceptable to Newton's contemporaries occasionally diverged from simpler options that were known to Newton, including the use of calculus and the acknowledgment of attraction as metaphysics. Newton's work was unique enough for Cohen, a translator of the *Principia*, to describe it as a signature method, or "style."[19] In Newton's style, subsequent steps to problems revise earlier steps, "as a result of comparing the conditions of the mathematical construct with the external world. . . . This alteration produces a new or revised stage one, a new and more complex mathematical construct."[20] This style implies a reading process that is both predictable in rhythm and recursive, echoing my discussion of the transactional nature of reading in this book's introduction.

The *Principia*'s metaphors ground abstract terms for readers briefly before the narrative proceeds to mathematical abstraction. Newton's examples prime readers to shift their minds easily from everyday, experimental phenomena to celestial events. Just like his mathematics would grow more complex as the *Principia* progressed, Newton's print descriptions facilitate the seamless movement between ideas. He describes an object we can easily imagine, "a spinning hoop, which has parts that by their cohesion continually draw one another back from rectilinear motions, does not cease to rotate, except insofar as it is retarded by the air. And larger bodies—planets and comets—preserve for a longer time both their progressive and their circular motions, which take place in spaces having less resistance."[21] In an experimental demonstration, a spinning hoop might make a discussion of planetary motion seem comically oversimplified. But when used as a metaphor instead of a prop, the spinning hoop becomes a tool that enables the visualization of hard-to-describe phenomena. In this context, the metaphors of the *Principia* build on the virtual witnessing model of the Royal Society, which Newton endeavored to use in his early optical publications. In the *Principia*, Newton does not pretend that assisting his readers with the visualization of certain phenomena is equivalent to replicating the phenomena themselves—as in a virtual experience. The spinning hoop, which demonstrates the permanence of God and eternal planetary motion, lacks a function beyond metaphorical instrumentality.

Even so, the attention of the *Principia*'s readers is directed virtually from the large and infinitely vast to the small and concrete. Newton's definitions inspire immediate transpositions from theories to concrete metaphor, rather than existing in the inaccessible realm of pure theory. As Newton discusses changes in density at the beginning of this section, he draws an immediate comparison to the differences in density between "snow" and "powder" ("Nive et Pulveribus") before returning to his mathematical content, writing that "the case is the same for snow and powders condensed by compression or liquefaction, and also for all bodies that are condensed in various ways by any causes whatsoever."[22] As in the example of the spinning hoop, metaphor leads immediately to abstraction, in this case a reference to the density of all bodies. The vocabulary used in the *Principia* is also straightforward. Following is an example from the scholium to the definitions at the beginning of the text: "quod moto loco movetur una locatum, adeoq; corpus, quod de loco moto movetur, participat etiam loci sui motum. Igitur motus omnes, qui de locis motis fiunt, sunt partes solummodo motuum integrorum et absolutorum."[23] Newton uses common terms such as "moto" and "loco"—movement and place—to demonstrate the universality of the terms he is defining.

The section of definitions also offers reader-friendly advice, including digressions like this one, where Newton addresses readers directly:

> I use interchangeably and indiscriminately words signifying attraction, impulse, or any sort of propensity toward a center, considering these forces not from a physical but only from a mathematical ("non physice sed Mathematice") point of view. Therefore, let the reader beware ("caveat lector") of thinking that by words of this kind I am anywhere defining a species or mode of action or a physical cause or reason, or that I am attributing forces in a true and physical sense to centers (which are mathematical points) if I happen to say that centers attract or that centers have forces.[24]

This passage furthers the normalization of mathematical extrapolation in that it brings attraction, impulse, and "propensity toward a center" together mathematically, thereby sidestepping religious or philosophical objections to the Newtonian world system. Beyond Newton's concerns that his definitions and attribution of causes be understood properly from a philosophical standpoint, this section also instructs readers, who learn to interpret his Latin word usage correctly. Newton's warning to the reader is another way of guiding the reader's attention in a way that produces a more correct reading of the text, and enhances the text's accessibility. From this section, the text proceeds to mathematical extrapolation, first by constructing a "mathematical framework of all the proofs, constructions, and exercises" of books 1 and 2 that give book 3 its foundation.[25] Newton encourages his reader to start thinking in terms of infinite series—an important bridge between the tangible and the intangible that buffets his use of mathematical extrapolation.

The method of perturbation further solidifies the relationship between the *Principia*'s form and content. This method was used in early work on planetary motion. Perturbation involves using a simpler, more knowable element to find better, approximate solutions to complex problems when ordinary methods yield incorrect results. Mathematicians take solved problems and then perturb or adjust them repeatedly in order to replicate the problem conditions and results they expect. Guicciardini, Michael Nauenberg, and Curtis Wilson have all analyzed Newton's use of perturbation methods in the *Principia*. Nauenberg describes Newton's use of the simplicity of binomial expansion in order to provide approximations.[26] Beyond the use of binomial expansion, it is also "extremely probable that in [Newton's] lost worksheets he must have integrated sinusoidal functions."[27] For Newton, analysis and geometry worked together in ways that were difficult to untangle.[28] Perturbation, then, while far from the only significant contribution of the *Principia*, was an essential part of its composition, and ensured its reception as a cohesive document that would help guide Newton's reception. Perturbation also reifies the *Principia*'s deployment of the uniformity of print. Using perturbations to generate mathematical solutions—one key to knowledge production in the *Principia*—means repeating a particular perturbation, which can be adjusted many times until the problem works. In the repeated disturbances perturbations cause, the solution gradually emerges by manipulating the problem itself rather than by removing the artificial element. When we see this artifice—or this art—yielding an approximation of truth in a prose monograph from 1687, it is crucial to stop and consider how using methods like perturbation helped to create Newton not only as a translator of truth but as a creator himself.[29] The repeated interjection of Newton's creativity into mathematical proofs, often at the *Principia*'s more challenging points, helps the book retain its narrative cohesion. This cohesion enhances the uniformity of the monograph and elevates Newton's status as its author.

The *Principia* and its economies of contingency are represented abstractly through the language of mathematics: print publication makes this abstraction archivable. A text like the *Principia* must of necessity be a kind of archive, to be consulted at length over time, but is also designed to be read with some degree of linearity.[30] Time is both absent and hyper-present in the *Principia* because of the *Principia*'s inert interface as well as the nature of its content, in which objects speed up, slow down, and stop, depending on the phenomenon illustrated. Those who translate this archive—in the form of popularizations—can never fully replicate reading the *Principia* as a timed experience but can still facilitate access to it and normalize its speed of acquisition for the reader.[31] The popularizations I describe later in the book manipulate the speed of readers' acquisition in their own ways.

Reading Calculus in the *Principia*

A brief reference to the origins of calculus in the *Principia* exemplifies the tiered accessibility that Newton incorporated into the *Principia*'s form: book 2, section 2, lemma 2. This passage became controversial later in Newton's career, during the calculus priority dispute with Leibniz.[32] Newton claimed he had performed all of the *Principia*'s calculations using calculus, but then had to redo them using traditional Euclidean geometry so that the book would fit discursive norms. He cited book 2, section 2, lemma 2 as a section of the *Principia* where his use of calculus lingered. Some scholars, like Guicciardini, agree, because some of the math in the *Principia* could really only be accomplished "by rather sophisticated integration techniques."[33] Cohen, however, disagrees with Newton's assertions that he had secretly proven many of the *Principia*'s formulae using calculus, in part because Newton chose to leave the *Principia* as it was for many years after its success, without rewriting it to include calculus.[34] Guicciardini mentions that Fontenelle's anonymous preface to l'Hôpital's *Analyse des infiniments petits* mentions the existence of calculus in the *Principia* and finds the text itself not inaccessible; its Latin may have made the book easier for speakers of Romance languages like Fontenelle to understand.[35]

Differential calculus involves analyzing the behavior of a particular section of a function at a point or range of points. For example, a point moving along a curve is accelerating or decelerating at a different rate at a different point, and calculus helps us understand these rates. Newton refers to calculus indirectly in this passage through using the concept of the "moment," in Latin *momentum*. "Momentum" is used here not to refer to force, but instead to refer to the behavior of a particular function at a particular point as it travels:

> *Momentum Genitæ æquatur momentis Terminorum singulorum generantium in eorundem laterum indices dignitatum et coefficientia continue ductis.*
>
> Genitam voco quantitatem omnem quæ ex Terminis quibuscunq; in Arithmetica per multiplicationem, divisionem et extractionem radicum; in geometria per inventionem vel contentorum et laterum, vel extremarum et mediarum proportionalium absq; additione et subductione generatur. Ejusmodi quantitates sunt Facti, Quoti, Radices, rectangula, quadrata, cubi, latera quadrata, latera cubica, et similes. Has quantitates, ut indeterminatas et instabiles, et quasi motu fluxuve perpetuo crescentes vel decres-

centes hic considero, et eorum incrementa vel decrementa momentanea sub nomine momentorum intelligo: ita ut incrementa pro momentis addititiis seu affirmativis, ac decrementa pro subductitiis seu negativis habeantur.[36]

This passage is a version of the product rule for finding derivatives, today also called Leibniz's Law. In spite of the paucity of calculus available in the *Principia,* an aggressive reader with an interest in mathematics could understand it from a reading of the foregoing section. In the product rule, one finds the derivative of a product of two quantities, what Newton calls "the moment of a generated quantity."[37] This derivative is found by adding together two quantities, each of which is the derivative of one function multiplied by the value of the other function. Thus, d(xy) = dx*y + dy*x. When Newton describes the "roots multiplied continually by the exponents of the powers of those roots and by their coefficients," he refers to the simplest form for taking the derivative of a function, where, given the function $3x^2$, the coefficient, 3, is multiplied by the 2 of the exponent and the exponent reduced by one.[38] The derivative in this example would then be 6x. Newton's description of roots being "multiplied continually" refers to the fact that derivatives refer to motion over an infinite range of points. When one calculates a derivative—in this case 6x—one shows that the formula is true for the function x at 2, 2.00001, 2.00002, and onward.

Even though the *Principia*'s mathematics became more difficult and challenging as it progressed, readers could understand many of its points superficially, and readers with some mathematical knowledge could have gleaned a rudimentary knowledge of what would become known as calculus. In this passage, the words would have made sense to a reader of Latin who had some background in mathematics.[39] This lemma's place in the less-read book 2 meant, however, that few readers would have encountered it. As they read through the passage, the varying use of the word "momentum" or "moment" may have confused readers slightly. In the passage, Newton first uses the term "momentum" or "moment" to describe the behavior of a function at a particular moment in time—as simple a term as he could have used to describe such a phenomenon. Later, however, he refines his definition of the term "moment" in significant ways: first, to include the way that functions increase or decrease at a given moment, such that a moment can be negative or positive. The concept of the moment, then, becomes confusing to readers as they proceed through this lemma.[40] Even though it is a little confusing, readers could still have understood that Newton was describing the behavior of a function at a particular point: "increasing or decreasing as if by a continual motion or flux; and it is in their instantaneous increments or decrements that I mean by the word 'moments.'"[41] Another point of confusion might be that a term associated with time, like a moment, could be negative, or the behavior of an object at a moment might be mistaken for the object itself. Even so, the fact that a rudimentary understanding of differential calculus could come from reading this section shows that even deep within the *Principia* Newton's usage retains transparency.

In spite of this glimpse at calculus, Newton had several reasons for using Euclidean geometry instead of calculus throughout the *Principia*. These reasons included the fact that the book's audience would be better able to understand the better-known Euclid than a new form of mathematics. If Newton were to combine a new mathematical system, such as calculus, with a new world system in which that mathematical system was essential to his arguments, an attack on each system could be used to undermine the other. However, the established geometry of the ancients could not be questioned publicly. Newton would even show multiple solutions to the same problems in an effort to prove veracity and to offer another window into his problem-solving methods.

As the second edition of the *Principia* (1713) makes clear, Newton also needed to be careful as he demonstrated universal gravitation. Cotes's preface to the second edition states, "[People] are always prattling on and on to the effect that gravity is something occult, and that occult causes are to be banished completely from philosophy."[42] Note that Cotes refers to these phenomena in relatively accessible language that could be translated to a variety of contexts apart from mathematics; here, in a book often thought of as inaccessible, one at least finds theoretical transparency. Cotes ascribes this superstition to followers of Cartesian natural philosophy, writing that "occult causes are the refuge of those who assign the governing of these motions to some sort of vortices of a certain matter utterly fictitious and completely imperceptible to the senses."[43] Newton did work with objects outside of immediate sense perception, but his union of celestial and terrestrial phenomena made distant objects seem familiar and less remote, rather than fetishized objects of mystery.[44]

Newton later lamented the decrease in classical knowledge among readers of the *Principia*. As Guicciardini writes, Newton was interested in "the mathematical competence of his readers and we will find him complaining about the fact that this competence had changed in the decades following

1687. According to Newton, most of the *Principia* was written in a style for 'philosophers steeped in geometry,' who by the early eighteenth century had disappeared."[45] By contrast, the readers of the second and third editions had more background in algebra and analytical mathematics. Common assumptions that the *Principia* was a book that bewildered most readers at the time of its publication, and then slowly became accepted, do not seem to align with Newton's own perspective on how the book itself was read and received by the audience he knew.

Visuality and Typography

In addition to uniting form and method, and creating a type of tiered accessibility within its narrative, Newton used typographical distinction and images to further the *Principia*'s rhetorical aims and contextualize the *Principia* in late seventeenth-century print culture. The title page to the 1687 first edition of the *Principia* is rendered in plain typeface and shows distinct differences from Hooke's *Micrographia,* which contains the Royal Society's crest and motto.[46] In Newton's text, the words *Philosophiae* and *Principia* are emphasized, evoking the title of Descartes's *Principia Philosophiae:* Newton's *Principia* presented a formal and mathematical response to Descartes's text. The dedicatory page in *Micrographia* sets off Hooke as a humble subject, using frequent typographical emphasis. Hooke's dedication reads, "I Do here most humbly lay this *small* Present at *Your Majesties* Royal feet. And though it comes accompany'd with two *disadvantages,* the *meanness* of the *Author,* and of the *Subject;* yet in both I am *incouraged* by the *greatness* of your *Mercy* and your *Knowledge.*"[47] Hooke thus opens *Micrographia,* a folio monograph with extensive treatment of microscopy and engravings that remain visually arresting today, by adhering to standards of politeness that require writers to feign insignificance such that they do not appear to threaten someone of high rank. Newton would, as we will see later, move in the opposite direction with the *Principia* and allow himself to be shown in an elevated rank.

Typography and *mise-en-page* assisted Newton with the presentation of content in the *Principia,* and helped to distinguish his work from works by others in the Royal Society like Hooke. As a point of contrast, consider how the beginning of each work is set off typographically. *Micrographia*'s text begins on a verso page, headed with three horizontal rows of cast fleurons, with an opulent drop-cap "A" beginning the text.[48] Twenty-eight words of the first paragraph are italicized for emphasis. Adjacent to this page is the book's first fold-out, which contains ornate engravings for several instruments, including a microscope with embossed decorations on its body tube. These engravings align with the overall message of *Micrographia* as a book that celebrates fine detail.

The *Principia*'s embodiment of mathematical elegance includes its visual presentation, its naked linear borders, and its simple line woodcuts. Its first sections, "Definitiones" (Definitions) and "Axiomata Sive Leges Motus" (Axioms, or The Laws of Motion), are each headed by two straight horizontal cast border lines and followed by a third line.[49] The first letter of each is an enlarged drop cap that contains no adornment. The definitions and laws are written in italic above their descriptions; there are no other italics in the subsequent paragraphs to distract from the laws or emphasize other language separate from the laws. The illustration on the recto page is of a parallelogram with a lone diagonal rather than an ornate fold-out engraving.[50] Other than an occasional shaded circle to represent a weight, abstract lines and circles illustrate the opening pages of the *Principia*. The preface to the reader in the *Principia*'s first edition also uses typography to solidify its themes and to emphasize words. This section comes before the poem introducing the text, so many readers would have encountered it. On the first page of this preface, the Latin words for "Mechanics" and "Geometry" are in roman letters while most of the rest of the passage is in italics. Eleven forms of "Mechanics" are set off in roman type, and nine forms of "Geometry" are set off in the same way. This is only on the recto side of the page. The verso contains less typographical distinction. Newton keeps the typography on the author's preface in the second edition. In this edition, the author's preface is after the prefatory poem.

The relative placement of image and text differ between the *Principia* and *Micrographia*. *Micrographia*'s images occupy their own fold-out pages rather than positions within the text. Other than the cast fleurons and embellished drop caps on some of the introductory pages, images are absent from *Micrographia*'s body text. The *Principia*'s illustrations are *in situ,* surrounded by Newton's writing, with the exception of a fold-out page illustrating the comet of 1680–81. Newton's visual integration of image and text allows for greater ease of reading. Rather than have to flip forward or backward in order to refer to an illustration—which, in the case of *Micrographia*, might

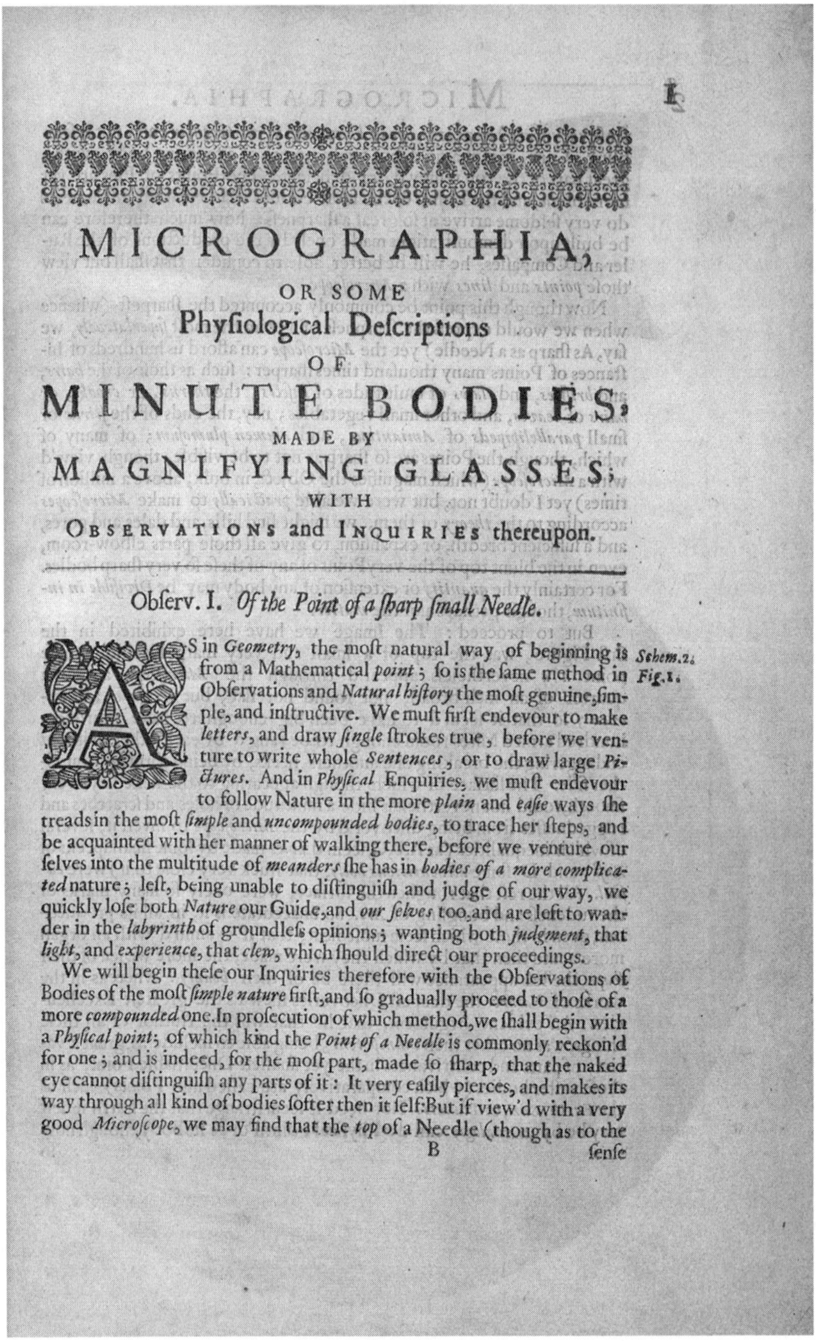

MICROGRAPHIA. 1

MICROGRAPHIA,

OR SOME

Phyſiological Deſcriptions

OF

MINUTE BODIES,

MADE BY

MAGNIFYING GLASSES;

WITH

OBSERVATIONS and INQUIRIES thereupon.

Obſerv. I. *Of the Point of a ſharp ſmall Needle.*

AS in *Geometry*, the moſt natural way of beginning is from a Mathematical *point*; ſo is the ſame method in Obſervations and *Natural hiſtory* the moſt genuine, ſimple, and inſtructive. We muſt firſt endevour to make *letters*, and draw *ſingle* ſtrokes true, before we venture to write whole *Sentences*, or to draw large *Pictures*. And in *Phyſical* Enquiries, we muſt endevour to follow Nature in the more *plain* and *eaſie* ways ſhe treads in the moſt *ſimple* and *uncompounded bodies*, to trace her ſteps, and be acquainted with her manner of walking there, before we venture our ſelves into the multitude of *meanders* ſhe has in *bodies of a more complicated* nature; leſt, being unable to diſtinguiſh and judge of our way, we quickly loſe both *Nature* our Guide, and *our ſelves* too, and are left to wander in the *labyrinth* of groundleſs opinions; wanting both *judgment*, that *light*, and *experience*, that *clew*, which ſhould direct our proceedings. *Schem.2. Fig.1.*

We will begin theſe our Inquiries therefore with the Obſervations of Bodies of the moſt *ſimple nature* firſt, and ſo gradually proceed to thoſe of a more *compounded* one. In proſecution of which method, we ſhall begin with a *Phyſical point*; of which kind the *Point of a Needle* is commonly reckon'd for one; and is indeed, for the moſt part, made ſo ſharp, that the naked eye cannot diſtinguiſh any parts of it: It very eaſily pierces, and makes its way through all kind of bodies ſofter then it ſelf: But if view'd with a very good *Microſcope*, we may find that the *top* of a Needle (though as to the

B ſenſe

FIGURE 1. First page, Robert Hooke, *Micrographia: or Some Physiological Descriptions of Minute Bodies Made by Magnifying Glasses, with Observations and Inquiries Thereupon* (London: Martyn and Allestry, 1665). (Wellcome Library, London)

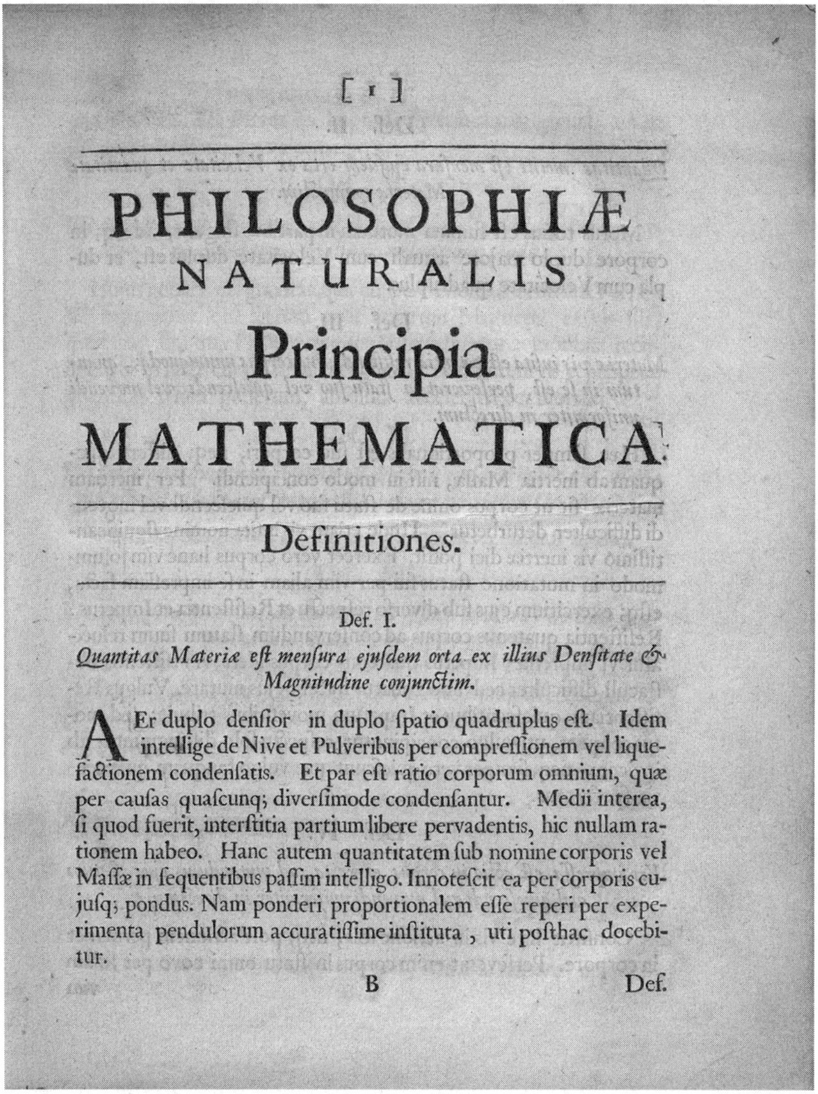
[1]

PHILOSOPHIÆ
NATURALIS
Principia
MATHEMATICA

Definitiones.

Def. I.

Quantitas Materiæ eſt menſura ejuſdem orta ex illius Denſitate & Magnitudine conjunctim.

AEr duplo denſior in duplo ſpatio quadruplus eſt. Idem intellige de Nive et Pulveribus per compreſſionem vel liquefactionem condenſatis. Et par eſt ratio corporum omnium, quæ per cauſas quaſcunq; diverſimode condenſantur. Medii interea, ſi quod fuerit, interſtitia partium libere pervadentis, hic nullam rationem habeo. Hanc autem quantitatem ſub nomine corporis vel Maſſæ in ſequentibus paſſim intelligo. Innoteſcit ea per corporis cujuſq; pondus. Nam ponderi proportionalem eſſe reperi per experimenta pendulorum accuratiſſime inſtituta, uti poſthac docebitur.

B Def.

FIGURE 2. First page, Isaac Newton, *Philosophiae Naturalis Principia Mathematica* (London: Joseph Streater, 1687). (Wellcome Library, London)

have to be folded out, then back in again—the *Principia* interlays word and image, creating a less visually symmetrical but more practical reading interface.

Newton's illustrations of pendulums and wheels rolling on globes, of abstract circular orbits, and of the moon's orbit are identical-looking circles and lines that represent vastly different types of object. Although the diagrams in the *Principia* are different kinds of object, they are not—following the introduction to John Bender and Michael Marrinan's *The Culture of Diagram*—arbitrarily selected. Bender and Marrinan define diagrams as "manifestly selective packets of dissimilar data correlated in an explicitly process-oriented array."[51] In the case of the *Principia*, the "explicit process" is mathematical extrapolation, and the "dissimilar data" of Newton's diverse examples are linked visually in addition to narratively and mathematically.

A pair of images from different books of the *Principia* demonstrates the role of Newton's diagrams in smoothing the narrative's movement outward from simple objects to complex systems. In proposition 4, theorem 4, from book 1 of the *Principia*, lines and circles represent abstract motion rather than specific objects. This proposition states that "the centripetal forces of bodies that describe different circles with uniform motion tend toward the center of those circles and are to one another as the squares of the arcs described in the same time divided by the radii of the circles." Each of the two circles is meant to represent the orbital path of a mythical object around a central point *S*, each orbit stabilized by centripetal force. The elements are described neutrally, as circles and bodies with radii.

As the *Principia* progresses, readers encounter problems with more bodies, with greater complexity and specificity. Later in the *Principia*, at the end of book 2, in the scholium to book 2, section 9, an illustration shows three circles that surround another point, which is also labeled *S*. These circles are described as "three orbits described about the sun S," and their orbits are first described, then compared to the movement of Mars in orbit. This time, the *S* at the center of the orbits explicitly refers to the sun. This specificity helps readers make sense of why previous points at the center of abstract orbits, such as that in book 1, proposition 4, theorem 4, were also named *S*, like a mystery that is unveiled at the end of a text. These two illustrations, placed next to each other, would present little difference among them without the language of Newton's text to provide context. Their similarities demonstrate the extent to which the formulae and examples from

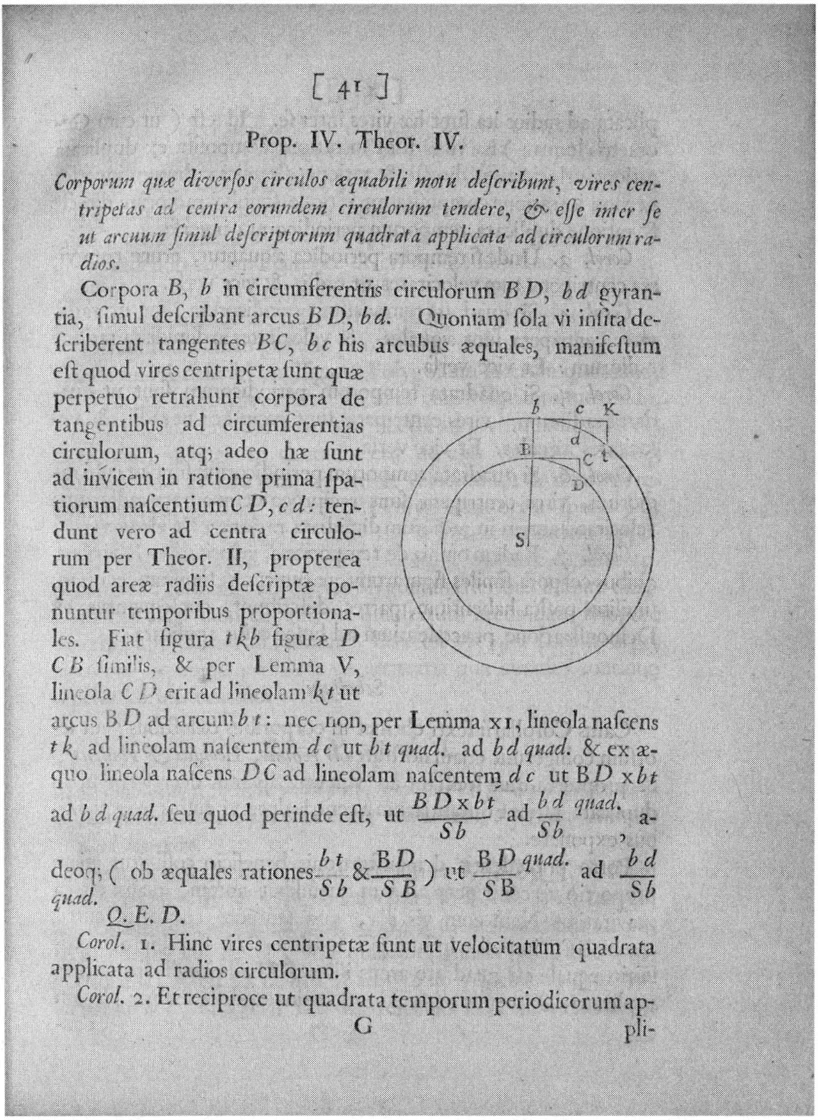

[41]

Prop. IV. Theor. IV.

Corporum quæ diverſos circulos æquabili motu deſcribunt, vires centripetas ad centra eorundem circulorum tendere, & eſſe inter ſe ut arcuum ſimul deſcriptorum quadrata applicata ad circulorum radios.

Corpora *B*, *b* in circumferentiis circulorum *B D*, *b d* gyrantia, ſimul deſcribant arcus *B D*, *b d*. Quoniam ſola vi inſita deſcriberent tangentes *B C*, *b c* his arcubus æquales, manifeſtum eſt quod vires centripetæ ſunt quæ perpetuo retrahunt corpora de tangentibus ad circumferentias circulorum, atq; adeo hæ ſunt ad invicem in ratione prima ſpatiorum naſcentium *C D*, *c d*: tendunt vero ad centra circulorum per Theor. II, propterea quod areæ radiis deſcriptæ ponuntur temporibus proportionales. Fiat figura *t k b* figuræ *D C B* ſimilis, & per Lemma V, lineola *C D* erit ad lineolam *k t* ut arcus B *D* ad arcum *b t*: nec non, per Lemma XI, lineola naſcens *t k* ad lineolam naſcentem *d c* ut *b t quad.* ad *b d quad.* & ex æquo lineola naſcens *D C* ad lineolam naſcentem *d c* ut B *D* x *b t* ad *b d quad.* ſeu quod perinde eſt, ut $\frac{BD \times bt}{Sb}$ ad $\frac{bd\ quad.}{Sb}$, adeoq; (ob æquales rationes $\frac{bt}{Sb}$ & $\frac{BD}{SB}$) ut $\frac{BD\ quad.}{SB}$ ad $\frac{bd\ quad.}{Sb}$

Q. E. D.

Corol. 1. Hinc vires centripetæ ſunt ut velocitatum quadrata applicata ad radios circulorum.

Corol. 2. Et reciproce ut quadrata temporum periodicorum ap-

G pli-

FIGURE 3. Proposition IV, theorem IV, Isaac Newton, *Philosophiae Naturalis Principia Mathematica* (London: Joseph Streater, 1687). (Wellcome Library, London)

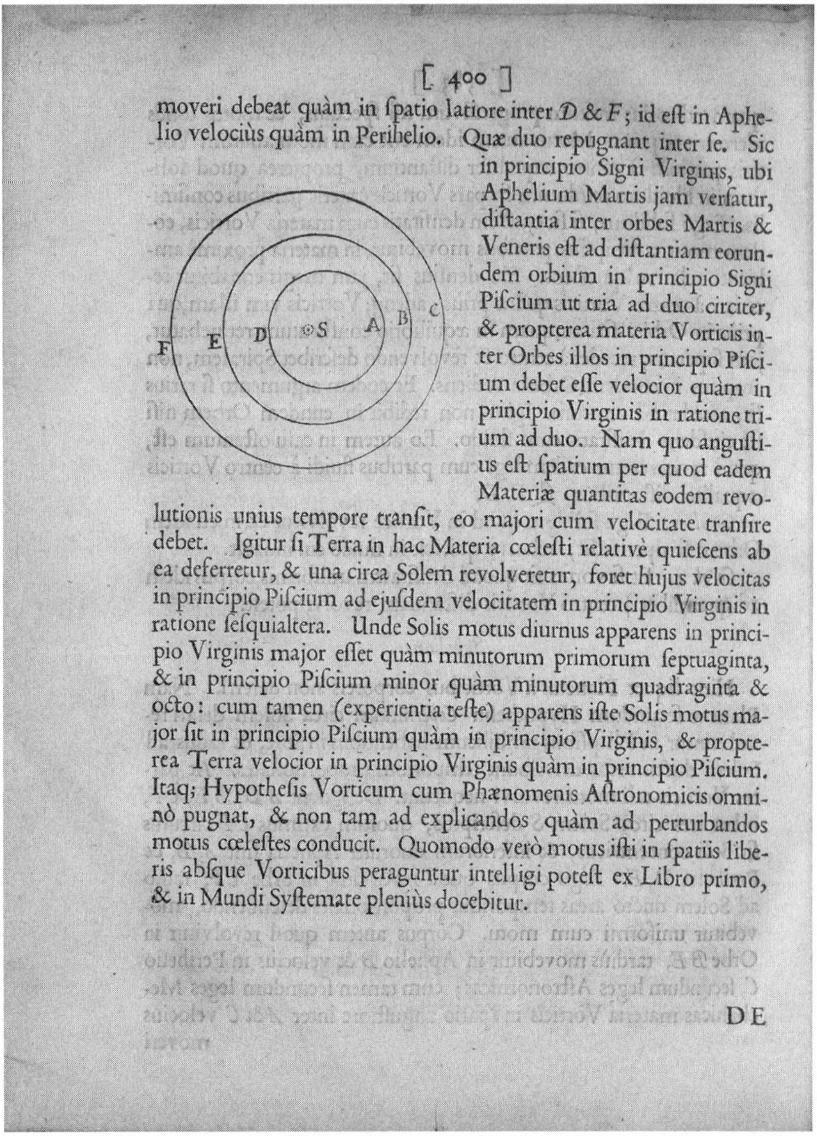

[400]

moveri debeat quàm in ſpatio latiore inter *D* & *F*; id eſt in Aphelio velociùs quàm in Perihelio. Quæ duo repugnant inter ſe. Sic in principio Signi Virginis, ubi Aphelium Martis jam verſatur, diſtantia inter orbes Martis & Veneris eſt ad diſtantiam eorundem orbium in principio Signi Piſcium ut tria ad duo circiter, & propterea materia Vorticis inter Orbes illos in principio Piſcium debet eſſe velocior quàm in principio Virginis in ratione trium ad duo. Nam quo anguſtius eſt ſpatium per quod eadem Materiæ quantitas eodem revolutionis unius tempore tranſit, eo majori cum velocitate tranſire debet. Igitur ſi Terra in hac Materia cœleſti relativè quieſcens ab ea deferretur, & una circa Solem revolveretur, foret hujus velocitas in principio Piſcium ad ejuſdem velocitatem in principio Virginis in ratione ſeſquialtera. Unde Solis motus diurnus apparens in principio Virginis major eſſet quàm minutorum primorum ſeptuaginta, & in principio Piſcium minor quàm minutorum quadraginta & octo: cum tamen (experientia teſte) apparens iſte Solis motus major ſit in principio Piſcium quàm in principio Virginis, & propterea Terra velocior in principio Virginis quàm in principio Piſcium. Itaq; Hypotheſis Vorticum cum Phænomenis Aſtronomicis omninò pugnat, & non tam ad explicandos quàm ad perturbandos motus cœleſtes conducit. Quomodo verò motus iſti in ſpatiis liberis abſque Vorticibus peraguntur intelligi poteſt ex Libro primo, & in Mundi Syſtemate pleniùs docebitur.

DE

Figure 4. Scholium to book 2, section 9, Isaac Newton, *Philosophiae Naturalis Principia Mathematica* (London: Joseph Streater, 1687). (Wellcome Library, London)

the *Principia* built upon one another in order to produce a cohesive and systemic narrative from simple to complex.

In addition to the less-intimate form of the *Principia,* its simplicity screened readers from the labors underlying its narrative's elegant construction. The *Principia*'s difficulty, its mathematically intricate language, and its impersonal tone may have blocked readers from virtually inhabiting Newton's mind, as one could do in the first-person narratives of the *Philosophical Transactions.* However, even though the *Principia* distanced readers from the labor that went into it, it also brought readers closer to the terrain that lay beyond it: closer to God and closer to the heavens. The *Philosophical Transactions* attempted to describe reality in its lively, first-person narratives of experiment; Newton's *Principia* presented the universe's phenomena with a formal seamlessness lacking in the experimental culture of the Royal Society.

Continental Context and Newtonian Form

Popularization does not happen in a vacuum, so to appropriately contextualize the prose of the *Principia* and the reception of Newton, we must consider the Continental monographs to which the *Principia* responded. Earlier monographs such as Descartes's *Principia Philosophiae* (*Principles of Philosophy,* 1644) and Galileo's *Dialogo sopra i due massimi sistemi del mondo* (*Dialogue on Two Chief World Systems,* 1632) influenced the form of the *Principia.* Newton owned Latin editions of Galileo's *Dialogue on Two Chief World Systems* and *Starry Messenger,* as well as several copies of Descartes's *Geometry,* a copy of his *Principia Philosophiae,* and a copy of his *Meditations on First Philosophy.*[52] Descartes's *Principia Philosophiae* was published in 1644, before the Royal Society was founded. Descartes is underexamined in recent accounts of seventeenth-century science in England, even though his works were published throughout the century, even into the 1690s.[53] Descartes's importance to Newton's work makes his presence in late seventeenth-century British scientific culture necessary to explore. Newton's text also extended the work Galileo had done in his *Dialogue on the Two Chief World Systems.* Galileo had argued that celestial and terrestrial forces were the same, but he had not made the mathematical proofs for this assertion. Newton derived the proofs and applied the same mathematical formulas universally, both to phenomena on earth and to celestial phenomena. Newton's Three Laws of Motion responded to Johannes Kepler's

Three Laws of Planetary Motion: if motion is the same everywhere, planetary motion does not require a separate set of governing laws.[54]

Newton could not, however, have used the formal characteristics of writers like Galileo in the *Principia*.[55] One problematic form for Newton was the dialogue, which Galileo had used successfully in multiple works, including his *Dialogue on Two Chief World Systems* and *Dialogues Concerning Two New Sciences*. Form was important to the reception of Galileo's work; his dialogues participated in a different classical tradition from Newton—that of the philosophical dialogue. The *Dialogue on Two Chief World Systems* had the initial title of *Dialogo* in large block letters across the top of its page, followed by Galileo's name, so its form and its author were the first two things one learned about it, rather than the words "principles" and "philosophy," which were emphasized in Newton's text. The *Dialogue* includes discussions between two philosophers, Salviati and Simplicio, and a layman, Sagredo, about the Copernican and Ptolemaic world systems, with the end result a conversion to the Copernican system.[56] The dialogues use the framework of reading explicitly by placing an ideal reader as the text's central target.[57] This character, Sagredo, begins as a neutral figure who lacks knowledge of either the Copernican or the Ptolemaic system; by the end of the text, he has been won over to the side of Salviati, who argues for the Copernican world system. Sagredo, the *Dialogue*'s ideal reader, is neutral, intelligent, and open-minded. Both Salviati and Sagredo, the heroes of the tale, were named for friends of Galileo. Simplicio, a weak intellectual villain, has his arguments knocked down repeatedly by Salviati and loses in the end. These characters are defined by their mental orientation toward philosophical dialogue—that is to say, by their ability to filter and analyze oral arguments. The governing structure of this form is human understanding and speech—reading *as* interpretation—rather than mathematics.[58] Using a neutral third party to cite himself was, for Galileo, a way to establish truth within a courtly environment, where indirect conversation was an effective rhetorical strategy.[59] Newton presented himself as separate from the social intricacies of contemporary natural philosophy, as exemplified by the *Principia*'s gestures to classical rather than contemporary contexts. For Newton to employ dialogues as Galileo did would align him too closely with the immediacy of the *Transactions*' first-person narrative.

Descartes's *Principia Philosophiae* (1644) and Newton's *Philosophiae Naturalis Principia Mathematica* also differ in form and in content. Descartes's *Principia* is composed of four parts, the first of which expands on the principles of human knowledge that Descartes had previously examined in his *Discours de la méthode* (*Discourse on the Method*, 1637). The second part aims to understand material objects and the laws of motion. The third part uses the theory of vortices—not universal gravitation—to explain celestial motion. The final part of Descartes's *Principia* contemplates the earth's origins. For Newton, these are both first and last principles. Descartes discusses such phenomena as heat, light, gravity, electricity, and magnetism and presents the first mechanical theory of magnetism. This organization makes sense according to the book's goal, which is to understand the principles of philosophy, broadly conceived. Considered independently of mathematical principles, philosophy is broad and encompasses the depths of human knowledge in addition to theories of the heavens and of the earth's origin.

Unlike Descartes's *Principia Philosophiae*, which begins with human thought, the opening pages of Newton's *Principia* seem to come from an unpeopled, undifferentiated world. Inertia, not thought, is the first concept Newton introduces.[60] Descartes emptied his mind until all that remained was the awareness of thought; Newton considers a universe where all is still until motion arrives to alter its course. From there, Newton presents laws of motion to the reader. After rest comes motion, then comes change in motion, then the calculation of changes in motion as equal and opposite. It is only the fourth and last section of Descartes's *Principia* that is "De Terra" or "About the Earth." Descartes addresses many similar topics to Newton, including "The Principles of Material Things."[61] These same principles are treated in Newton's *Principia*, and Newton would not have considered them secondary to human thought. For both writers, the narrative structure reflects the methodology used: Descartes worked deductively and Newton inductively. The *Principia* presented a comprehensive system of the world to supersede Descartes's system, Aristotle's system, and many other competing systems of the world. Yet the universality of Newton's images in comparison to Descartes's is striking. Descartes also shows orbits; however, to illustrate his points regarding smaller bodies, he uses a hand holding a ball in a sling, forcing the reader to confront the differences between the earth and the heavens, the abstract and the real. For Newton, the effacement of visual difference among images helps to reveal, rather than conceal, the truth. The spare design of the *Principia* reflects its connection to divine authority: as a reflection of God's universe and the mathematical laws governing it, it requires no excessive decoration from its writer or printer.

Newton's and Descartes's texts contrast not only visually but also in principle and organization. The primary principles under discussion in Descartes's *Principia* are intellectual principles, not physical principles. If Newton may exemplify the shift to modern science as a discipline, one could point to his change in title as a key component of that shift because of its increased specialization and its explicit mention of mathematics.[62] Descartes's *Principia* is laid out like a prose narrative, where a small roman numeral heads each paragraph, describing its contents rather like the headings in a Bible. These headings show that the readers of Descartes's *Principia* were expected to open the text and read it in different places, as opposed to Newton's *Principia*, where they were expected to build upon concepts from axiom to axiom, especially in books 1 and 3.[63]

Enhanced Paratexts across Different Editions

The *Principia*'s paratexts further eased its reception because they made the book easier for readers to navigate. Newton's explicit discussion of the *Principia*'s print strategies occurs in the front matter, which was comparatively straightforward and understandable to an audience that had some Latin knowledge. The *Principia*'s form and its paratexts distinguish it from the other works that compose the context of its publication: it does not begin by jumping immediately into calculations, as Newton's own correspondence and the *Philosophical Transactions* often do; instead, it begins in all three Latin editions with a poem. For the first edition, Edmond Halley wrote an ode introducing the *Principia*, an ode that was reproduced in its entirety in the subsequent Latin editions of the *Principia*, and quoted in the first English-language edition.[64] The poem elevates the work and hails the system that Newton has witnessed and recorded before its readers have witnessed it for themselves, defining its author and his status as superhuman. Dedicatory odes were fairly common in monographs of the sixteenth and seventeenth centuries. A choice to write a book like the *Principia* meant that such praises could be heaped on its author through a preface.

Prefatory materials make explicit and implicit statements about how a book and its writer should be received. The publication of a landmark book by a member of the Royal Society was an important event, even if its writer was an irritable outsider like Newton.[65] Genette, whose work on paratexts—broadly described as the materials outside of the body text of a book—informs this section, writes of prefatory materials that "when an author is anxious to highlight his merit, talent, or genius, he generally prefers—not unreasonably—to entrust this task to someone else by way of an allographic preface, sometimes a highly suspect one."[66] Although Newton did not beg Halley to depict him as a conqueror of heaven, Halley's poetic preface functions as a lightning rod for the *Principia,* adding charge and energy to its claims. Newton was no pirate, but the Newton of the poem "opens the treasure chest of hidden truth," cavorting with the gods who have given him unique access to divine wisdom.[67] Newton may not have written the poem, but he also wrote nothing to contradict the effects of this flash of lightning on the *Principia:* he remained reticent, allowing readers to draw their own conclusions regarding his work's divine connections.[68]

Newton's own preface to the *Principia,* while more subdued than Halley's poem, is equally strategic in design, from its dedications, to its discussion of the Royal Society, to its explanation of the *Principia*'s publication as a monograph. Newton does not address the preface to the Royal Society, even though the Society gave the book its imprimatur; nor does he address the preface to James II, the *Principia*'s dedicatee. Instead, Newton responds to the ancients, praising their respect for mechanics and geometry as means of understanding phenomena that occurred beyond the reach of experiment. His choice to sidestep the concerns of the dominant scientific culture of his community shows the extent of Newton's attempt to resist that community's pull. Thus, the preface following the poem demonstrates the book's goal to lay out the mathematical principles of natural philosophy, and then "to demonstrate the other phenomena from these forces" largely through mathematical extrapolation.[69]

In 1713 Newton added several elements to the *Principia*'s second edition that increased its accessibility to readers, further easing his reception. Subsequent editions of the *Principia* progressed to greater accessibility overall, including the addition of both a table of contents and an index in the second and third editions. These apparatuses assisted readers in finding information that was part of a particular category. In the case of a very difficult book, an index and table of contents gave readers additional entrances for understanding. Its index includes entries such as "attractio corporum universorum demonstratur III, 7," and "attractionis causam vel modum auctor nusquam definit 6,2: 160, 18: 88,12: 530, 2."[70] The phrase "attractio corporum universorum demonstrator"—"the attraction of all bodies is demonstrated"—is a relatively straightforward assessment of the *Principia*'s thesis. The latter phrase refers to the lack of definite causes for attraction, or the occult force of gravity—important concepts for the popularization of the *Principia.* We take accessible textual apparatuses for granted today, but it is significant that the *Principia*'s essential ideas were relatively accessible in two of its three editions. To pursue the inaccessibility argument is to ignore the many handles the book's readers were given across its editions, and to privilege the ability to reproduce math over the ability to understand content.[71]

The General Scholium was an addition to the second edition that concluded the *Principia* with a moving narrative of how the rules proven by Newton connect to God's creation and divine authority. The most frequently repeated words in the General Scholium are the varying Latin forms of the word "God." Newton writes that God "is not eternity and infinity, but eternal and infinite; he is not duration and space, but he endures and is present. He endures always and is present everywhere, and by existing always and everywhere he constitutes duration and space. Since each and every particle of space is *always*, and each and every indivisible moment of duration is *everywhere*, certainly the maker of all things will not be *never* or *nowhere*."[72] In this passage, Newton relies on readers' received ideas about an existing and ever-present deity in order to connect the God they know to the seemingly abstract principles Newton has just presented in the *Principia*. In this manner, Newton could be conciliatory toward the Fellows of the Royal Society and at the same time remain reverent toward God. Newton refers to a living God who is not a reified icon but who is eternally present throughout the world and the heavens. As such, God has been a part of the *Principia* all along, because his existence surrounds us: in the permanence of matter; in universal gravitation; and as a perpetual enduring presence. Rather than argue that God is behind Newton's principles, Newton instead shows readers that the concepts they associate with God—concepts such as infinity, permanence, and presence—are consistent with the *Principia*'s mathematical principles, from its definition of absolute concepts such as time, space, and matter, to its universal applications of physical principles. The General Scholium was also one of the more easily understood sections of the *Principia*, so its distillation of the book's central tenets helped explain Newton's ideas to readers who could not understand them.

Other paratextual material in the *Principia* gestured toward its reading public. Following the General Scholium to the 1726 third edition was a common paratext: a list of other books offered for sale by the publisher, a "Catalogus Librorum Prostantium apud Guil & John Innys" in which twenty-seven works were advertised. These books were all current, recent editions and included both Newtonian works and original works of scholarship. Some were written in English and some in Latin. Editions of the *Philosophical Transactions* and *Opticks*—popular works written in the vernacular—were also listed. The English edition of the *Opticks* was listed first, before the Latin edition, and before a quarto of Newton's *Analysis per Quantitatum Series* (1711). This implies that the publishers assumed at least some of their audience would be interested in reading books in the vernacular; the list's other entries reflect the assumption that not all of the public's reading interest would be in original scientific or Newtonian works. The list advertises popularizations by s'Gravesande, *The Religious Philosopher* by Cartesian Bernard Nieuwentyt, and several religious works by John Ray. The last three works on the list are even more afield from Newton and Newtonianism, including works on architectural style and "new memoirs of literature, containing an account of new books printed both at home and abroad; with dissertations upon several subjects, miscellaneous observations, etc. for February 1726." From this paratext, we can see that the *Principia* itself advertises popularizations and works with accessible content that appeal to nonspecialist readers.

The paratexts to subsequent editions revealed increasing transparency about the *Principia*'s subject matter, so there was a formal interest in broadening its public reach. Editorial practices and public interest appear to inform one another here, as they would later do in Newtonian popularizations. Roger Cotes, Plumian Professor of Astronomy and Experimental Philosophy at Cambridge, and a protégé of Newton's, wrote the editor's preface to the 1713 edition. Cotes's preface summarizes the most important findings of the *Principia* and makes Newton's work more accessible to those whose with a limited ability to understand the *Principia*. Cotes begins his preface by emphasizing the importance of extrapolation to the *Principia*. He writes that the text begins with "what is simplest and nearest to us . . . the nature of gravity . . . in terrestrial bodies, so that when we come to celestial bodies, so very far removed from us, we may proceed more securely."[73] With these lines, Cotes supports my claims regarding the *Principia*'s intelligibility: Newton's text begins simply and understandably, using the examples of objects unaffected by resistant media, then leads readers into difficult material, guided by the security of what they have read in the previous pages.[74] This preface, less figurative than Halley's poem, nonetheless echoes its idea that Newton is a divine figure with a unique access to truth.

Cotes's preface bridges the poem's content to the content of the text, as well as to the divine references that Newton would include at the end of the *Principia,* in the General Scholium. Cotes writes that "he must be blind who does not at once see, from the best, and wisest structures of things, the infinite wisdom and goodness of their almighty creator; and he must be mad who refuses to acknowledge them."[75] If readers were unable to imagine the mathematics that went into making Newton's arguments, they could more easily grasp the concept of the book: to reveal, then demonstrate, the mathematical complexity and soundness of the universe—the sensorium of God. The presence of an editorial preface strategizes the use of print practices and is helpfully picked up again thematically at the conclusion of the book, in the General Scholium. The inclusion of a preface by someone other than Newton controls the distance—and, therefore, the attraction—between Newton and his readers. Newton's, and his editors', ability to theorize complexity, describe attraction, and use perturbation theory to artificially structure problems were translated to these readers in divergent ways, at different distances.

Newton's editorial changes to the *Principia*'s title page help to explain the way Newton viewed himself as an author, and how he desired to be viewed by others. One of Newton's copies of the *Principia*'s first edition, housed at the Wren Library, contains changes Newton made for the second edition, including a correction to the author's name on the title page. Instead of leaving "JS Newton" as printed, he spells his name out as "Isaaco Newtono." This desire to spell out his name shows that Newton thought carefully about the presentation of the *Principia*'s front matter to readers. This editorial change differs from the phenomenon observed in I. Bernard Cohen's essay that examines the title page of Newton's *Opticks.*[76] The first edition of the *Opticks* was published in 1704, nine years before the second edition of the *Principia.* Cohen proposes several reasons for the decision to leave Newton's name off of the title page, among which are the likelihood that others knew who Newton was anyway, and that peer scientists such as Christiaan Huygens had published similar books without their names on the title page. The Newton who wanted to be known for the all-encompassing *Principia* may have simply wished to memorialize himself. The self-presentation of Newton in print—whose authority was more explicitly linked with the *Principia* than with the *Opticks*—carried over into visual culture. In the portrait of Newton painted by the studio of Enoch Seeman (1726) in London's National Portrait Gallery, the text Newton is reading is the *Principia,* and the globe behind him is a celestial globe. The *Opticks* sits closed at the edge of his desk, and optical instruments are absent from the painting. To

FIGURE 5. *Sir Isaac Newton,* Studio of Enoch Seeman, c. 1726–1730. (National Portrait Gallery, London)

the left, behind Newton, is a bookcase filled with large bound volumes, most likely in folio and quarto, showing Newton's status as a learned man and his use of other printed works in his own research.

Other editorial changes to the different editions preserved intelligibility for the reader. The organization of the beginning of book 3—an essential section for understanding Newton's System of the World—shows how editorial practice and transparency worked together. Book 3 of the first edition begins with a recap of what has preceded it, followed by a list of nine hypotheses.[77] Below hypothesis 1, we find "[N]atura enim simplex est et rerum causis superfluis non luxuriat."[78] This translates to "Nature is simple and does not luxuriate in superfluous causes," an essential tenet of understanding the universe as mathematically representable. Hypothesis 4 is "centrum systematis mundane quiescere," that "the center of the system of the world is at rest," a hypothesis that links back to inertia, the first law of motion.[79] Book 3 of the second edition changes "hypotheses" to the more assertive "regulae philosophandi" or "philosophical rules," followed by "phaenomena" and "propositiones."[80] This reorganizing and renaming helped keep the book organized thematically, and alerted readers to the presentation of new conceptual information. The second edition also added a guide at the side of each page to let readers know which book they are reading. In the third book, the recto page says "liber tertius" and the verso says "de mundi systemate." The third edition (1726) keeps this layout, adding a fourth rule to the "regulae philosophandi," but keeping the organizational principles intact.[81]

Beyond his interest in adding the more accessible General Scholium and other useful paratexts to the *Principia*'s second and third editions, Newton's decision to publish three Latin editions of the *Principia* had diverse origins. One practical reason for publishing second and third editions of the text is stated in Cotes's preface to the second edition: copies of the first edition were becoming rare and expensive to the readers who wished to own them. Newton also wished to amend some of the proofs with better solutions, a major reason to put out a new edition of any book. He also wanted to make sure the work remained in copyright after the Statue of Anne (1709–10). Another—more political—reason is hinted at in the dedicatory pages to the three editions, as each is dedicated to a different monarch: James II, Anne, and George I. Charles II also receives credit as the founder of the Royal Society. Newton, who would become Master of the Mint and president of the Royal Society, connected his own successes to four different monarchies with vastly different politics. This choice may have made Newton appear to be outside of political concerns. As I show in later chapters, Newton would ultimately be seen as a representative of all that was good about England: the breadth of these dedications implies that Newton's loyalties to England transcended historical and political specificity, just as his research attempted to do—and just as the subsequent popularizations of Newtonian natural philosophy would do.[82]

Early Reception

Newton's union of narrative and method simultaneously connected and separated the small objects readers could touch and those far beyond their reach. This attraction and distance affected part of the book's early reception. Guicciardini has analyzed how some readers read the *Principia*, with a focus on expert readers such as Newton himself and his contemporaries Christiaan Huygens and Gottfried Wilhelm von Leibniz. These readers offer essential historical records of their interactions with Newton's work as ideal, not common, readers. Guicciardini writes that "the number of people competent enough to follow Newton's demonstrations was, and still is, very limited. [There is a] lack of sufficient information. Of course, absence of evidence is not to be confused with evidence of absence."[83] The ideal readers of the *Principia* are studied because of their importance in the history of science, philosophy, and mathematics. Because of their expertise and sophisticated understanding, their recorded encounters with the *Principia* are extensive. However, a scarcity of material evidence for common readers of the *Principia* does not mean it cannot be part of the history of reading: the *Principia* itself offers clues to how it was designed to be read.

Not all readers, however, needed to master the *Principia*'s content. It would have been sufficient for some readers to understand its main arguments, with the opportunity to pursue those arguments further if desired. Many in its audience read the *Principia* differently than people would read a comprehensive science text like it today. As Patricia Fara observes, "Written in Latin and packed with geometrical diagrams, the *Principia* appears a dry book, but for those who understood it, Newton wrote in a persuasive style. Right at the beginning, he stated his three laws of motion, which govern how objects move and interact with one another."[84] Most readers would not have been examined on the *Principia*, though many would be expected to produce some type of response to this "persuasive style." They might address Newton's ideas in a relatively intimate forum, such as correspondence or conversation, or they might acknowledge his work more formally, perhaps by incorporating or responding to the *Principia*'s points in their own research. Few readers understood—or would be expected to understand—the entire text. Newton knew this. Newton's correspondence directed readers explicitly: "When you have read the first 60 pages, pass on to the third book & when you see the design of that you may turn back to such propositions as you shall have a desire to know, or peruse the whole in order if you think fit."[85] These same sections of the *Principia*—book 1, sections 1–3, and book 3—became its most commonly taught sections in university curricula during the eighteenth century.[86] Newton casually mentions the first sixty pages in his letter as if it is a round number, but page 60 is precisely where section 3 of book 1 ended. Iliffe describes these directions as meant for "an intermediate group of readers," as "a means of getting through the *Principia* in a reasonable amount of time," and these instructions were, as I have shown, complemented by Newton's writing and format.[87]

Correspondence between Newton and one of the *Principia*'s early readers demonstrates how a representative reader responded to the *Principia*. Gilbert Clerke, a mathematician and fellow of Sidney Sussex College, Cambridge, from 1648 to 1655, wrote to Newton in 1687, asking for some clarifications after having read the *Principia*. Clerke writes, "I confess I do not as yet well understand so much as your first three sections, for which you do not require that a man should be *mathematicè doctus;* but if I must not tell you, till I understand those sections and your third book very well, perhaps I must never tell you, your book coming to me in the evening of my declining age."[88] In these lines Clerke admits that he, a fellow mathematician, has difficulty following even the simpler parts of Newton's book. Even though he did not have the time to devote to a thorough understanding of the *Principia* due to his advanced age (he left his position at Sidney Sussex thirty years prior), Clerke nonetheless remained interested enough in the book to write to Newton. In his letter, Clerke mentions some problems with proposition 17 in book 1 of the *Principia,* where Newton describes the calculation of a body's path, given the presence of centripetal force and a constant velocity.[89] Clerke, it must be noted, understood the *Principia* enough to ask questions about it. Newton's response demonstrated courtesy and tolerance toward Clerke's incomplete understanding: "I do not wonder that in reading a hard book you meet with some scruples and hope that the removal of those you propound may help you to understand it more easily."[90] Newton did not expect each reader, even those who had some mathematical skills, to understand a "hard book" like the *Principia*. Yet the retired mathematician Clerke still wanted to work toward understanding it.

Clerke knew that he could never understand the *Principia* fully, but he wrote several more letters to Newton about the book's points, remarking that "had your book come to me 20 or 30 years ago, I would have taken more pains than now I can; but now I have begun I will try a little to get some general notions about the planets and tides."[91] Even though Clerke, trained as a mathematician, believed that it would take him years to get through the *Principia*, he was still interested in a rudimentary knowledge of Newton's work, and he possessed enough knowledge to understand the *Principia*'s overall points. Clerke, an avid and knowledgeable reader, was shown the limits of his own understanding through encountering Newton's challenging text.

Another early reader, John Craig, a former student of Newton's at Trinity College and himself a mathematician, also found the *Principia* enticing, if distancing. Craig wrote the following to his Edinburgh acquaintance, the mathematician Colin Campbell:

> I have delayed hitherto that I might send you some satisfactory account of Mr. Newton's book, which I found to be a task of no small trouble; and though I have not yet perused one quarter of it, and have an unsatiable desire to know the whole; yet knowing how extremely acceptable it will be unto you I have sent you the book itself, which I hope will abundantly apologize for my not writing to you sooner.
>
> [PS] Be not hasty in reading over Mr. Newton's book for I shall let you have it as long as you please.[92]

Craig had begun to read the *Principia,* but, like Clerke, he found it difficult and doubted he would understand all of it. Even so, Craig's interest in the book bound him to the *Principia* longer than he would prefer. Craig delayed in writing his friend Campbell because he had fallen behind in his reading. Although both men were mathematicians, the *Principia* challenged them. In this letter, the *Principia* serves a dual function: it excuses a lack of correspondence and it substitutes for that correspondence. Craig does not produce the promised précis of the *Principia* for his friend; rather, he produces the book to speak for itself. Sending along the *Principia* by way of apology for not having written sooner, Craig lets Newton's text represent his own intellectual limitations. To send the book may have been a tacit plea for Campbell's assistance in understanding Newton's text further. Like Clerke, in spite of the frustrations of reading the *Principia,* Craig nonetheless finds an "insatiable desire" to fully understand the book.

Readers like Clerke read the *Principia* knowing that they would never understand all of it, contacting Newton himself for clarification. Others, like Craig, might use the *Principia*'s difficulty as a means of bonding with other intellectuals. Yet they remained interested in its inaccessible qualities even though Newton's text revealed their insufficient understanding. It is readers like Craig and Clerke who I believe exemplify the *Principia*'s audience, because they did have some training in mathematics such that they could read parts of it. The *Principia* itself encourages this range of familiarity, with its accessible front matter, its incremental increases in difficulty, and its interplay of detailed, often very challenging, demonstrations with universal concepts, as well as its consistent visual elements and diagrams.

In the introduction to the *Principia*, Newton addresses its complexity and difficulty, perhaps anticipating Clerke's letters and those of other mathematicians and natural philosophers, "I earnestly ask that everything be read with an open mind and that the defects in a subject so difficult may be not so much reprehended as investigated, and kindly supplemented, by new endeavors of my readers."[93] According to this introduction, the *Principia*'s difficulty is a perversely friendly invitation, one meant to inspire communication and dialogue among those who do not understand it fully. These lines also imply that the *Principia* represents a beginning, initiating new research into the connections between mathematics and natural philosophy.

Years after the publication of the *Principia*, one writer who cited the centrality of its form and organization was Daniel Defoe, whose *A System of Magic* (1726) Al Coppola analyzes in *The Theater of Experiment*—noting especially the manipulation of the title and title page as a gesture to Newtonian natural philosophy. Coppola's book focuses largely on experimental culture, but this reference to print supports my claims about the extent to which print facilitated popular Newtonianism. Coppola contends that Defoe's satirical reference to Newtonian science "titillat[es] the reader with the playful suggestion that the book might be a how-to manual for black magic," inspired by the "law-giving and system-building" of Newton's *Principia*.[94] If Defoe was mindful enough of the *Principia* to satirize its print context in 1726, the same year that the *Principia*'s third edition was published, before it had even been translated into English, then we may conclude that other writers associated the print presentation of Newtonian natural philosophy with its reception.

The *Principia* sorted its audience into various levels of intellectual understanding, where Newton sat almost at the top, next to the God whose works he celebrated.[95] Considering the text's readership, this kind of ranking was especially controversial: many of the scholars and intellectuals across Europe who read the *Principia* (including other Fellows of the Royal Society) would have been considered Newton's peers or even his superiors, not his inferiors. Newton was the Lucasian Professor of Mathematics at Trinity College, Cambridge, near Clerke's alma mater, Sidney Sussex College. As a distinguished member of an intellectual community, Newton was still part of that community and thus placed himself at risk by publishing such a comprehensive system and not participating in community publishing practices. Strikingly, Newton achieved the opposite effect, drawing his readers in by distancing them. Later, in 1752, Frederick the Great of Prussia would use Newton's solitary work and aloof status to justify absolutism, writing that "just as it would have been impossible for Newton to arrive at his system of attractions if he had worked in harness with Leibniz and Descartes, so a system of politics cannot be arrived at and continued if it has not sprung from a single brain."[96] These lines not only argue against the pressures of community as a kind of shackle, but also reinforce the distance between the ruler and his subjects and center governmental authority where Newton had located scientific authority.

I conclude by describing some "readers" who may have never read or even intended to read the *Principia*, especially in its later editions. Owning a copy of it as a material object was a symbolic act, and there were presentation copies that bear few traces of readership. If a book was given as a presentation copy, there is often little evidence of actual readership, in part because the purpose of the gift is not to read the book, but to brand it as a worthy prize for academic accomplishments. Trinity College, Dublin, gave a gold-tooled, gilt-edged copy of the *Principia*'s third edition to Francis Gore for taking second at Hilary term (1735). David Gregory gave another copy of the third edition to Lewis Bagot (c. 1764), who became a clergyman. By midcentury, the *Principia* was significant enough as an object, regardless of who read it, to be presented as an honored prize in elaborate binding—the recipient endowed with the symbolic knowledge that he possessed the universe in a book.

The *Principia* was not isolated among scientific treatises—it responded to other seventeenth-century natural philosophical texts, yet diverged from these works in form and narrative because it appertained to a late seventeenth-century print context when journal publication had begun to predominate.[97] Newton's ability to conjoin paratext, form, structure, illustrations, typography, and method—many of which were choices unavailable to publishers of individual articles—created a printed book with visual and narrative coherence.[98] With the *Principia*, Newton did not simply write a challenging, fascinating treatise: the material conventions of the monograph united with mathematical method to reveal to readers the limits of their understanding. Its cultural significance, shaped by the medium and circumstances of its initial publication, would flourish immediately following Newton's death in 1727. The next chapter, which takes us from the 1687 poem opening the *Principia* to James Thomson's 1727 poem celebrating Newton, examines the complex figuring of Newton's masculinity and orientation to a print community in each.

3 Newtonian Popularization and Masculinity in Two Poems about Newton

This chapter and subsequent chapters expose the indistinct boundaries between the popular and the professional in the transmission of Newtonian ideas. We will now traverse the gap between the *Principia* and popular culture through analysis of popular representations of Newton in poetry. Poetic works about Newton and his ideas, which circulated widely, are often overlooked yet essential to understanding popular Newtonianism. Poetry helped transmit Newtonian ideas to the public through a gendered lens—one inspired by early representations of Newton. Newton's gender had first been invoked explicitly in the *Principia*, especially through the poem Edmond Halley wrote as part of its front matter. Later poetry about Newton then refined and explored his masculinity further, teaching the public to associate scientific excellence with superior masculinity. As Patricia Fara writes, readers' knowledge of Newton came "from chatty journal articles or reading poetry, then an enormously popular genre for transmitting ideas."[1]

Printed works adopted a variety of approaches to science and poetry. In her work on emerging disciplinarity, Robin Valenza writes that divisions within disciplines separate professions; they "divide[] the physicist's work from the poet's."[2] The distances between physics and poetry are evident in printed works, though in many ways these distances are contrived and manipulated, especially in the case of popular Newtonianism. The positionality of the poet and the scientist varies vis-à-vis knowledge, creativity, and readership. I build on the kind of work Fara and Valenza have done to show

how the print practices of the *Principia* inspired early popularizations of Newton during a time when disciplinarity was less concrete, when the poet and the physicist commented on their common territory of print authorship. To write popular Newtonian poetry is to stake a claim to authorial as well as aesthetic territory. The poems' discussions of authorship show that gender—specifically the framing of male authorship—was an important component of the transmission of Newton through poetry.

There are several critical precedents for discussing the issues of publicity, privacy, and gender that surface in my analysis of Newton in poetry. Publication was a gendered action as well as an intellectual one, and scientific publication was a predominantly male arena. Jürgen Habermas identifies the emerging bourgeois gentleman as a new type of male figure—one whose domestic life is directed outward toward the public sphere.[3] Philip Carter and others have written about the ways men circulated in enclosed public spaces like theaters and coffeehouses.[4] As Shapin and Schaffer write, late seventeenth-century philosophers were pressured to be sociable—not just to royal patrons, but also to one another.[5] One's public reputation could become tied to one's gentlemanly status, a tension between public and private we have seen emerge in Newton's reading notes and in the publication of the *Principia*. The publication of a work like the *Principia*—a work that is simultaneously innovative, deserving of public commendation, and the product of private contemplation—troubled the lines demarcating public and private.

Wendy Wall's analysis of print and masculinity in the Renaissance—what it meant to be a "man in print"—has implications for discussions of masculinity and authorship in later cultures of print.[6] Her proposal that "gender is an issue at the level of the commodity," specifically the print commodity, is especially useful to understanding Newton as a masculine figure whose sales depended largely on his unique authorial reputation and on print publication.[7] Even publications by interested women of the seventeenth century, including Margaret Cavendish's *A Description of a New-World, Called the Blazing-World* (1666), acknowledge an imbalance between male and female authorship. Susan Lanser's recent work on "sapphic subjects" demonstrates that "the sapphic became a flash-point for epistemic upheavals that threatened to dismantle" an order that aligns with representations of Newtonian masculinity.[8] Newton's print practices commented on his masculinity—he chose to separate himself from the contemporary, "gentlemanly" scientific community. This separation meant that his gender also conformed less to that community's standards of judgment, in which the timing and tone of debate, conversation, and the dissemination of knowledge had been normalized. Newton used mathematical extrapolation to reveal how the universe was both incomprehensibly distant and yet mathematically definable; this method, combined with its fixity in a print monograph, meant that its author was correspondingly unique. Establishing this separateness from other men, which included Newton's distance from their conversation, was a key component of Newton's popularity and popularization: as a result of this separateness, he was portrayed as a superior, masculine figure in imaginative literature. To read poetry about Newton was to be presented with perspectives on gender and science. Other intellectuals wondered whether Newton was a man like other men, or whether there was some unique quality that appeared—like a Benjaminian aura—when meeting Newton: all of these speculations were linked to print.[9]

James Thomson's "A Poem Sacred to the Memory of Sir Isaac Newton" (1728) was published at a time Blanford Parker describes as a transitional one for Augustan poetics, citing among other examples the decline of classicism in Thomson's work.[10] Parker writes that "it is probably an anachronism to attribute the turn to literalism . . . to an interest in science. [Cowley] and Thomson . . . embraced science, and their work found a solid center in contemporary physical theory."[11] While I agree that the eighteenth century is far more complex than a neat imagined map of literary "literalism," empiricism, and scientific interest, scientific culture was an inextricable part of eighteenth-century culture, and some familiarity with scientific topics should have been expected of imaginative writers of the period, in particular those who were active socially in London. Many easy-to-find, inexpensive works integrated popular print and science—including *the Athenian Mercury, the Ladies' Diary*, and the *Gentleman's Magazine.*

The Royal Society helped popularize scientific practices with publications like the *Philosophical Transactions* as well as with public demonstrations of scientific experiments. Yet the Society was also an association of gentlemen, many of whom socialized in other settings. Newton was interested in experimental philosophy, and he researched astronomy and mathematics, where it could be an asset to work independently. Figures like Samuel Pepys and Robert Boyle were often held up for their embodiment of the Society's gentlemanly ideals in a culture that privileged witnessed experimentation, but Newton was less sociable, preferring the geographic isolation of Cambridge, and the social isolation of working independently. He also disliked publication in the *Philosophical Transactions* because the style privileged first-person prose articles, the essays were comparatively short, and the journal was fast-paced. D. Christopher Gabbard shows that in Aphra Behn's *The Dutch Lover* (1673), masculinity existed along a spectrum of machismo, effeminacy, and an English middle ground in which Englishmen were strong and masculine, yet also civilized.[12] Robert Iliffe refers to the scientist's idealized masculinity when pointing out Newton's "self-fashioning" in print and the idolatry that Newton's astronomical treatise engendered; I see Iliffe's work as opening rather than closing the conversation about Newton and masculinity, a conversation that has rich potential for literary study.[13]

Both Halley (1656–1742) and Thomson (1700–1748) celebrated Newton's presence in social and scientific communities and, importantly, connected sociability and science to masculinity, as embodied in print publication. The two writers use the concepts Newton worked with in the *Principia* as inspiration for poetry, especially the concepts of attraction and extrapolation. Both poets likewise center on the medium of print as it was conceived during their respective times, as if in order to write about Newton, one must also make claims about print publication. This cements the importance of Newton as a print figure that was earlier developed through the publication and construction of the *Principia*. The early eighteenth century was an important time for print authorship, as Manushag N. Powell has shown, including "the unsettling possibilities that an explosion of authorship brings."[14] This unsettlement was a scientific one as well as a social one. Thomson, who, I show, borrowed heavily from Halley's Latin ode introducing the *Principia*, reveals changes in the writer's relationship to the print marketplace, especially the way print fashions masculinity against an implied community. Both print communities and scientific communities had changed by the 1720s and 1730s. Written early in his career, Thomson's "A Poem Sacred to the Memory of Sir Isaac Newton" helped to popularize Thomson's own works as well as Newtonian ideas; it even contributed to the success of his landmark work, *The Seasons*.

1687: Newton, the Man Who Showed Us the Promised Land

Newton chose to publish the *Principia* as a monograph rather than as an article or series of articles in part because of his ambivalence about print.[15] As I mentioned in the introduction, Newton wrote to Robert Hooke (1635–1703), curator of experiments at the Royal Society and author of *Micrographia* (1666), that "there is nothing which I desire to avoid in matters of Philosophy more than contention, nor any kind of contention more than one in print. What's done before many witnesses is seldom without some further concern than that for truth."[16] As a partly audience-driven medium, print could indeed be corrupt when it came to the transmission of specialized, verifiable knowledge.[17] The popularity of shorter works like the *Transactions* fostered this kind of contention because the shorter length and shorter time to publication resembled a conversation.[18] When Newton published the *Principia,* he manipulated print and narrative conventions in such a manner that a running dialogue was nearly impossible.[19]

The *Philosophical Transactions* had aligned with the ideas of gentlemanly sociability and the vibrant intellectual culture of the Restoration. The *Principia* was a different kind of book. It even had an "ungentlemanly," less-conversational tone because it was written in Latin and focused on mathematics more than on experiments that could be witnessed in a community. Furthermore, its comprehensibility relied, as the previous chapter has shown, on textual apparatus and form. The *Principia* may have even verged on the fantastic, because Newton had offered a workable and thorough explanation of the entire universe, condensed in one ordinary-looking text. Newton's introduction addressed itself to the ancients as well as to his contemporaries, claiming that both mathematics and mechanics, which the ancients "considered . . . to be of the greatest importance," composed the framework for his argument.[20] Halley's prefatory ode helped explain the position of Newton relative to the scientific community and to his gender. Its importance to Newtonian popularization cannot be overestimated: a simple Latin poem—a form its readers would have certainly found familiar—was far more likely to be understood by lay readers than the latter sections of the *Principia.*

The ode was not the only text Halley wrote to further Newton's interests. Before the *Principia*'s publication, Halley wrote an unattributed review that appeared in the *Philosophical Transactions.*[21] As with most reviews, Halley's was published in the back section of the *Transactions,* following the unlikely companion pieces of an account of a hermaphrodite and a review of a published natural history of plants.[22] Halley writes, "This incomparable author having at length been prevailed upon to appear in public, has in this treatise given a notable instance of the extent of the powers of the mind; and has at once shown what are the principles of natural philosophy, and so far derived from them their consequences, that he seems to have exhausted his argument, and left little to be done by those that shall succeed him."[23] The Newton described in these lines is averse to publicity, comprehensive in his methods, of an exceptionally powerful mind, and "incomparable." Halley's review characterizes Newton as someone who metaphorically stops time because he leaves so "little to be done" by others. The stoppage of time implied in these lines may also connect to the *Principia*'s controlled pace, with Newton's lucid Latin prose and its interconnected mathematical calculations, versus the more chaotic *Transactions*. The *Principia* begins with the simple principles of the three laws of motion before it proceeds to more remote celestial mechanics. All of its points were linked, which leaves the impression of a tightly woven text. By emphasizing the permanency and incomparability of Newton's accomplishments, Halley challenges those who will respond to Newton; Newton, having "exhausted his argument," leaves no room for literal or figurative response. This depiction of Newton's work as superior would eventually be translated into the portrayal of Newton's superior masculinity.

Halley's poem is most frequently described as an ode, yet its full title, "Ode on This Splendid Ornament of Our Time and Our Nation, the Mathematico-Physical Treatise by the Eminent Isaac Newton," is worth noting because it shows the slippage between Newton and print that also appeared in Halley's book review.[24] This title claims that the ode celebrates the book itself, a "splendid ornament" of England in the 1680s, produced by the "eminent" Newton. The ode invokes Newton as a semi-divine figure, claiming that "no closer to the gods can any mortal rise," and readers of the poem would be encouraged to think of Newton himself as a splendid thinker and an ornament to his country.[25] This brings to mind Courtney Weiss Smith's work on Newtonianism and religion, including the compatibility of religious and scientific narratives wherein "Newtonians eagerly and frequently affirmed God's constant activity in his clockwork."[26] In this instance, Newton is connected to the gods of classical mythology, but the divine connections still apply. Readers may have been confused by the placement of Newton's name in all capital letters, when the ode is dedicated to the book itself and not its author. The fact that the poem celebrates both the man and the physical book at once furthers the idea that the audience should see them as either interchangeable or yoked together. Making the claim that Newton *was* the *Principia* implies that those who did not publish their research or who published only in the *Philosophical Transactions* were somehow smaller than Newton physically, and therefore lesser men. A person can be joined to a book as its author, but only a journal's editor can represent its corpus.

The poem occupies a single page, front and back. Sandwiched between the author's preface and the definitions that open the *Principia*, the poem is located in a prominent site in the book.[27] Despite its placement, it has been underrepresented in discussions of the *Principia*'s—and Newton's—reception. Much has been made, even by Newton, of the fact that most people who read the *Principia* did not understand it. An anecdote related by Newton shortly after its publication described an undergraduate who claimed "there goes the man hath writ a book that neither he nor any one else understands."[28] My discussion of readers in the previous chapter also illustrates this limited reception qua reception, but I have reserved my analysis of a piece that people with some understanding of Latin were likely to grasp—Halley's ode—for this chapter on poetry.

The language of Halley's ode positions Newton in three main ways: in relation to other men, to other thinkers, and to other means of disseminating information.[29] At times these three kinds of representation combine in one example, as in the poem's extended comparison of Newton to Moses. Halley describes Moses, to whom Newton is compared favorably, as a transmitter of information more than as a liberator of the Jews: "He [Moses] who commanded us by written tablets to abstain from murder, / Thefts, adultery, and the crime of bearing false witness."[30] Describing Moses as having commanded "us" casts the poet and his audience as a community of readers, foregrounding the role of dissemination and reception in the text that will follow. Halley's comparison relates more to Moses's role in the dissemination of divine information than to Moses's place in the Old Testament. The word "scriptis," or "written," describes Moses's tablets rather than "inscribo," or inscribed. According to the poem, God dictated and Moses wrote, instead of God inscribing the tablets and Moses carrying them. Moses is both writer and transmitter, yet is inferior to Newton in both capacities, as Moses is listed among those who "did less than our author for the condition of mankind."[31] To compare Newton to Moses and minimize the accomplishments of Moses seems surprising, but Halley does so in the poem as context for his praise of Newton. The lines that follow, "Or he who taught nomadic peoples to build walled cities, or he who enriched the / nations with the gift of Ceres," also relate to the dissemination of information.[32] Here, engineering and agriculture are inferior to the *Principia* because the information Newton transmits is not "concerned with only a few / comforts of a wretched life," but is timeless and divine, "admit[ting] us to the banquets of the gods."[33] This statement evokes the culture of print to which Newton responded, including texts like the *Transactions*, which published smaller-scale observations on comparable topics like engineering and agriculture. Halley's poem even makes a coded reference to the superiority of print over manuscript in its reference to the first person who, "with a reed from the Nile," "set spoken words before the eyes."[34] This anonymous inventor of manuscript communication is, likewise, rated inferior to the author of the printed *Principia*, manuscript itself listed among Halley's list of "wretched comforts."

The framing of Newton's expertise in Halley's poem contrasts to other seventeenth-century representations of scientific inquiry. The Royal Society's aims and its members had been celebrated since its inception in the 1660s, but writers who worked on the topic of scientific achievement wrote more frequently about Francis Bacon, the Society's methodological ancestor. Abraham Cowley's (1618–67) poem that opens Thomas Sprat's *History of the Royal Society* had been an invitation to participate in natural philosophy, beckoning its audience to "Come, enter, all that will."[35] Cowley, writing before Newton had published in the *Transactions*, had also compared Bacon to Moses, as two participants in philosophical inquiry. For Cowley, Bacon and Moses had limits, because "Life did never to one Man allow / Time to Discover Worlds, and Conquer too."[36] These lines ally with the modes of Baconian inquiry the Society favored: this openness to many extends to the Society's inspirations as well as its members. If one person's life is not enough for full inquiry into natural philosophy, then the project of the Society is a communal one. A generation later, Halley proclaims that Newton—who has conquered worlds *through* describing how they work—is an exception to this rule, and also compares him to Moses. A human life span may be limited in terms of literal travel around the world to gather evidence and understand all philosophical principles from personal experience. However, that same life span, facilitated by improved instrumentation and flourishing print markets, allowed Newton time for metaphorical conquests and discoveries, to derive his own set of divine laws.

Moses, a figure from long before print, is still brought into the *Principia*'s conversation about Newton and information. Halley's description of Moses addresses information culture without specifically invoking print. Three of the commandments Halley selects in the passage—lying, theft, and adultery (in the form of adulteration)—relate indirectly to publication in the late seventeenth century. The word Halley uses for theft, "furta," implies duplicity in word as well as the theft of objects. "Adulteria," the word chosen to describe adultery, refers not only to adultery in the literal sense but also to contamination and alteration to make something less pure.[37] Impurity further suggests the potential contamination present in experimental science, versus the possibilities for purity in Newtonian mathematical precision. Lying is referred to by its full biblical treatment: the act of bearing false witness, which also evokes the use of witnessed experiment in contemporary experimental culture. Halley reminds readers that acts of witnessing are fallible. Newton personally witnesses divine phenomena—correctly—elsewhere in the poem, which furthers the importance of true and false witnessing. The fourth commandment Halley mentions, "caedes," Latin for "murder," implies striking a blow or cutting something off. The word is related to "caelator," meaning carver or engraver, and also connects, in a more extreme fashion, to adulteration and change. Establishing this context for Newton's witnessing diverges from Baconian observation: the difference between recording sensory perception and bearing a claim about the divine is a vast one.

The commandments Halley fails to mention include coveting goods or a spouse, honoring one's parents, keeping the Sabbath, taking the Lord's name in vain, and not placing gods before the Judeo-Christian God. Several of these seem irrelevant to a poem about science, not least because they were incompatible with scientific culture and it would be imprudent to call attention to them. Honoring one's literal father and mother was, of course, appropriate, but the late scientific revolution was also a time when one's intellectual ancestors drove the revision and refinement of scientific ideas, which often meant proving a "father" wrong. Following the poem's claims about Newton's superior masculinity, this connection to filial deference would further undermine its aims. Keeping the Sabbath was a commandment, but Fellows of the Royal Society, including Samuel Pepys, often conducted business or worked on accounts from home on Sunday.[38] The commandments depicted in the poem were considered important to Halley as a representative of late seventeenth-century science and print culture, and the majority of these relate to false communication and deceit as foils to Newton's achievements. The poetry of scientific innovation was thus tied to the consciousness of print as a potent medium and the awareness of scientific communities as a matrix for reception.

The beginning of the poem addresses the *Principia*'s reception by demanding the attention of an implied scientific community, asking readers to "behold the pattern of the heavens, and the balances of the divine structure."[39] Halley addresses an audience that is expected to regard a commonplace sight—the night sky—with unique attention, not as a jumble of stars, but as a system with a pattern and a balance that only Newton has identified. The community Cowley had celebrated in his poem about the Royal Society writes at the level of its readers, revealing observations without linguistic adornment. Cowley's philosopher's "candid Stile like a clean Stream does slide, / And his bright Fancy all the way / Does like the Sun-shine in it play."[40] The prose style Cowley describes was very different from the "Newtonian Style" that Cohen identifies, described in the previous chapter. As a comparison, Halley's audience is told to look up at the night sky, rather than down at a stream, for a variety of reasons. For one, the book makes important advances in celestial mechanics, so this opening frames the subject of Newton's text. Readers' understanding of the poem's opening is assisted with a knowledge of the original Latin in which it was written, including the description of Jove's "leges," or laws, in the opening lines.[41] The verb "to read" is also connected to the word for law. "Lego" means "I read," and "leges" means "you will read," so the idea of reading surfaces through this reference to law. The person reading the *Principia* would also read Newton's "leges motus," or "laws of motion"; the still attention to which Halley brings readers prefigures the introduction of inertia as the first law of motion. Describing these divine laws in the poem prepares the reader for Newton's content. Indeed, the poem's content resembles a fugue that the full *Principia* expands upon and develops; the poem's still beginnings lead to Newton's accomplishments, and its conclusion—Newton among the Gods—predicts the second edition's General Scholium. Newton is also positioned relative to a community through the use of the first-person plural "discimus," or "we learn," opposite him.[42] Cowley's poem does anticipate a later reference by Newton, who would later describe the sun playing on water, but this occurred when Newton discussed the refraction of light in the *Opticks*. These descriptions would lack the "bright Fancy" of the philosophers whom Cowley celebrates. Halley's representation rejects the lutic elements present in Cowley's poem because his subject stood in contrast to the dominant community.

The opening lines of Halley's ode direct the reader's attention to the heavens as a way to draw on shared knowledge of Newton's achievements:

Newton's prior renown came from his invention of the reflecting telescope in the 1660s.[43] Directing people to look up also means directing them to set aside their other pursuits and focus their attention on this new world system. The opening also asks for humility, because printing a mathematical system of the world, as Newton has done, is a divine project that requires intense personal focus and textual attention in order to be understood. The *Principia* thus requires a withdrawal from sociability to read. The poetic elevation of its author is also necessary in order for the significance of its truths to be gleaned.

The second stanza commands *"Surgite Mortales!"* ("Arise, mortals!").[44] The group whose attention has been invoked in the first stanza is now named here. They are not fellows of the Royal Society, but are bowing mortals compared to Newton's immortal achievement. Halley's poem questions the Royal Society's priorities in other ways, especially regarding mathematical bases for truth claims. Shapin writes that "notions of truth traditionally connected with mathematical culture were deemed out of place in experimental natural philosophy . . . as violations of epistemological decorum."[45] Halley argues the converse of this idea: experimental natural philosophy could distract natural philosophers from mathematical truths. A person who is focused on experiment may not connect it to mathematical principles. "The things that so often vexed the minds of the ancient philosophers / And fruitlessly disturb the schools with noisy debate / We see right before our eyes, since mathematics drives away the cloud."[46] The positioning of public and scholarly debate as raucous "noise" undermines the value of public standards for truth-making.[47] Others' opinions make it difficult to perform work accurately; by this measure, sociability, masculinity, and truth are not in harmony. Newton responded to the "vexing noise" of the Royal Society and its *Philosophical Transactions* by publishing the *Principia*. According to the poem, mathematics has the power to remove the clouds surrounding people's judgment.[48] These lines also suggest the literal clouds present when prior observers saw the sky solely as eyewitnesses, versus Newton's use of mathematical formulae to understand planetary motion—regardless of the weather.

If mathematics drives away the cloud, then the poem casts Newton as the masculine presence at the helm. A later passage discusses the way the *Principia* accounts for apparent contradictions to prior theories in astronomy.[49] Halley emphasizes the safety the audience may be sure of, now that some anomalous, feminized entities of the cosmos have been explained. "We learn at last why silvery Phoebe moves at an unequal pace, / Why till now, she has refused to be bridled by the numbers of any astronomer."[50] In this formulation, Newton holds the reins; he is the masculine astronomer who tames the moon, a gesture of domination that implies forced consent. Later, Halley envisions Newton's mathematical understanding of the tides to exemplify the scientist's even greater understanding of the moon. "We learn also the magnitude of the forces with which wandering Cynthia / impels the ebbing sea, While its weary waves leave the seaweed far behind / And the sea bares the sands that sailors fear."[51] Newton's mathematical calculations reveal the intimate rhythms of the moon and tides. These exaggerated comparisons show how Halley invents a poetic persona for Newton where he is cast as male, presented as a foil to a community. Although Newton had annotated the tale of Diana and Actaeon in Ovid's *Metamorphoses*, it was with an eye to peripheral details like the names of the hounds, rather than the voyeurism and power reversals at its center. Newton was not a physical conqueror; however, the far-reaching geographical implications of his work, coupled with the poem's invocation of classical mythology, become bedfellows nonetheless.

The Newton of this poem has accomplished the hypermasculine feats of regulating Phoebe and itemizing Cynthia's wanderings. Even though the poem is only talking about one entity, the moon, Newton is able to compel two metaphorical female bodies to submit to him, making him even manlier. The fact that he has allayed the fears of otherwise masculine men, the sailors, further shows Newton's superior masculinity relative to his own community, the Royal Society, who were frequently depicted as voyagers in pursuit of natural knowledge.[52] The *Principia*'s advances in astronomy provide the ingredients for fictionalizing Newton as a dominating, seductive male figure. What other kind of person could prove universal gravitation mathematically, and so thoroughly comprehend near and distant attractions? However, Newton's excessive social aloofness deviated from masculine norms that were present in late seventeenth- and early eighteenth-century Britain. He had to be considered apart from other men, like sailors and his peers.

The representations of masculinity in Halley's poem originate in the *Principia* itself. Newton uses modified examples of sailors and sailing ships in the Scholium to the Definitions to illustrate mathematical points, smoothing transitions between complex ideas with concrete examples: "if . . . a ship is being borne westward by sails and wind with a velocity of 10 units, and a sailor is walking on the ship toward the east with a velocity of one unit, then the sailor will be moving truly and absolutely in unmoving space toward the east with a velocity of 10,001 units and relatively on the earth toward the west."[53] Newton's narrative positions the sailor as very small—his relative motion 10,000 units slower than absolute motion; even so, the image of the sailor helps readers to envision the calculations Newton has performed. He valued the example of the sailor, because he refers back to it a few pages later to support a claim regarding absolute versus relative motion: "as in the above-mentioned example of the sailor."[54] Even the use of a sailor as an image makes a claim about masculinity and the *Principia:* sailors were simultaneously hailed for their masculinity and, during a time of dangerous ocean travel, known for their vulnerability to shipwreck as much as their mobility. Newton's calm manipulation of the sailor figure from afar reinforces his superior status as a mathematician and as a man.

Newton, Halley's poem argues, opens heaven to mortals in a way that no scientists have done. This accomplishment, an exact explanation of the system of the world, supersedes that of even revered figures like Moses. Halley effectively defends the *Principia*'s difficulty as commensurate with its divine connections, questioning the faith of readers who would challenge the authority of the printed book, Newton's authority, and, consequently, the authority of God. He concludes with the claim that "no closer to the gods can any mortal rise," toasting Newton with the nectar of the gods. He links Newton with the muses, a connection that sustained well beyond the end of Newton's and Halley's lives.[55] The poetry of scientific innovation was thus tied to consciousness of print as a medium and to awareness of scientific communities.

1728: Thomson's Masculine Print Community

The 1720s saw a shift in the reception of printed works. David Brewer describes this trend as "imaginative expansion," in which readers could turn fiction into "an opportunity for communal play, rather than partisan strife."[56] Brewer details the extent to which characters developed long imaginative afterlives among British readers and writers, using the publication date of *Gulliver's Travels* in 1726 as a starting point for this shift. Newton's death in 1727 coincided with this time period in literature, and works written about Newton had their own rich cultural afterlives. Courtney Weiss Smith's recent *Empiricist Devotions* works with Newtonian poetry, including Thomson's "Poem Sacred to the Memory of Sir Isaac Newton" (1728) and Desaguliers's *The Newtonian System of the World* (1728), in order to evaluate the connections between popular Newtonianism, empiricism, and religion. Smith's analysis foregrounds the importance of "seeing God in things" in Newtonian poetry as a means of understanding "constant divine action in the world."[57] Desaguliers's poem is not only one in which, according to Smith, "scientific knowledge leads to political truth."[58] Desaguliers also sets up a male community in which the world's kings (plural) yield way first to corruption and then to an understanding of the Newtonian system, followed by the crowning of Newton himself as a king surrounded by princes. The most striking aspect of Desaguliers's poem is its invasive print apparatus, in which footnotes take over the majority of the page space: the overt authority of Newton and the primacy of the printed medium intertwine. Thomson's poem furthers the conversation Smith identifies about Newton and divinity, finding Newton to be deeply linked with community and with print, as my analysis shows.[59]

To begin with, Thomson's poem plays an important role in cultural history. Following its initial appearance, Thomson's poem honoring Newton appears at the end of the 1730 edition of *The Seasons*. *The Seasons* was one of the most popular books of the eighteenth century, and "was said to be found among shepherds who had previously never seen any other book but the Bible."[60] It would have been present in far more homes than the *Principia*, and its popularity persisted until the nineteenth century. John Akin's *An Essay on the Application of Natural History to Poetry* (1777) compares *The Seasons* favorably to Virgil's Georgics.[61] Akin writes later that in *The Seasons* "we are to look for the greatest variety of genuine observations in natural history," so readers from the latter part of the century may have even chosen to read *The Seasons* based on its links to natural history. Even readers of *The Seasons* who may not have had much interest in science to begin with would have learned much about Newton from the poem.

The Seasons is also a highly important work in print history; the rights to publish it were part of two important copyright decisions, *Millar v. Taylor* and *Donaldson v. Becket*. Its presence in these cases reflects the popularity of *The Seasons* long after it was published and after the death of Thomson.[62] Newton was even mentioned in *Donaldson v. Becket*, so he is further connected to *The Seasons*. Lord Camden's speech regarding the reversal of the decree railed against unscrupulous booksellers. Camden idealized the publishing of elite works, using Newton as an example: "Glory is the Reward of Science, and those who deserve it, scorn all meaner views. . . . It was not for Gain, that Bacon, Newton, Milton, Locke, instructed and delighted the World: it would be unworthy such men to traffic with a dirty bookseller for so much as a Sheet of Letter-press."[63] This speech conflates writing for profit and illicit trade, elevating Newton and his illustrious peers as clean and above proprietorship, whereas we have seen that Newton's affiliation with Streater meant he was not above economic concerns. Camden's statement imagined distance between Thomson as a writer and Newton as a writer, affiliating Thomson with proprietorship, which would be the shape of things to come, and Newton with the pursuit of intellectual glory. And yet, Newton's name was actually used to sell *The Seasons*.

Thomson's poem on Newton had a far wider audience than has been previously assumed. The poem influenced the reception of *The Seasons*, because publishers took advantage of Newton's prestige when marketing it. In the 1730 295-page edition published by Millan and Millar, the Newton poem occupies prime advertising space at the center of the title page; in fact, Newton's name is printed in the same large type as Thomson's name. The words "Seasons" and "Hymn" are both eight millimeters high, and the word "Poem" in the title of the poem to Newton is one centimeter tall. This design was clearly intended to appeal to readers: the poem was written by the less-known Thomson during a time shortly after Newton's death when there was a thriving market for works memorializing Newton. The rubricated words of the title are "Seasons," "Hymn," "Poem," "Isaac Newton," and "Britannia," so both "Poem" and "Isaac Newton" would have held a reader's attention in bold red. The title's words themselves present some ambiguity about what they mean. Presented in linear form, the title page reads *The Seasons, a Hymn, a Poem to the Memory of Sir Isaac Newton, and Britannia, a Poem*. "Newton" and "Britannia" are the most specific and therefore vivid entities mentioned in this title. Because of this prominence, if a reader were to parse this title to decide what the book was about, the reader might reasonably conclude that the poem to Newton was as long and ambitious as *The Seasons*.

A 348-page edition (also Millan and Millar) from the same year contains the Newton poem but does not advertise it. A 1735 edition printed for Millan and Millar used Newton's name to help sell books and showed that Millar was still using Newton to advertise *The Seasons*. The listing of books printed for Millar in the back matter of this 1735 edition includes a mention of *The Seasons* and "A Poem to the Memory of Sir Isaac Newton" as well as

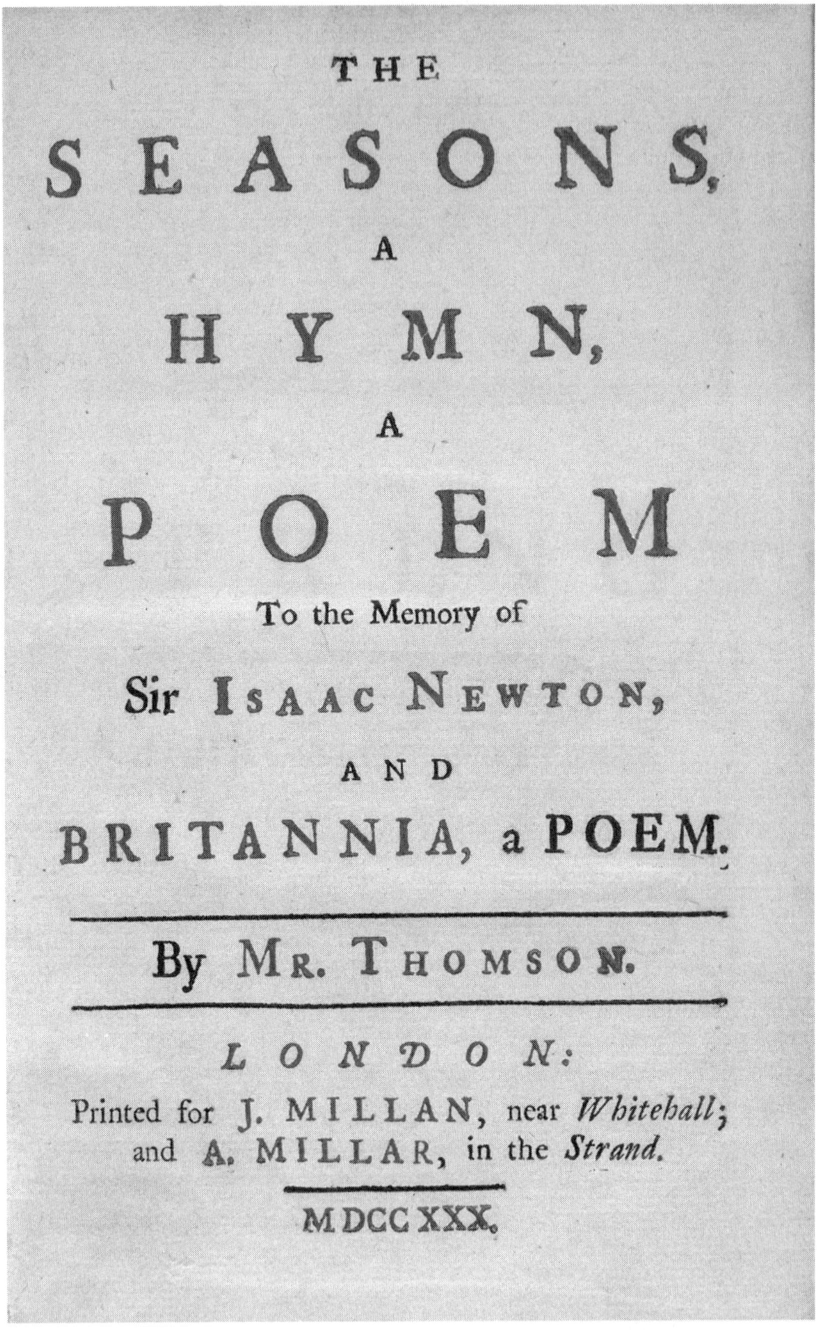

THE

SEASONS,

A

HYMN,

A

POEM

To the Memory of

Sir ISAAC NEWTON,

AND

BRITANNIA, a POEM.

By MR. THOMSON.

LONDON:

Printed for J. MILLAN, near *Whitehall*;
and A. MILLAR, in the *Strand*.

MDCCXXX.

FIGURE 6. Title page, James Thomson, *The Seasons, a Hymn, a Poem to the Memory of Sir Isaac Newton, and Britannia, a Poem* (London: Millan and Millar, 1730). (The Huntington Library, San Marino, California; call number 24382)

"Britannia." In the three-line advertisement, the longer name of the poem about Newton stands out more than the far-longer poem *The Seasons*. Even though the work about Newton was a short poem, it helped sell books. A 1794 Cooke's edition of Thomson has the poem about Newton as well as an "Explanation of the obsolete words used" in Thomson's poetry, so by this point Thomson himself requires some cultural translation much as Newton had inspired popularizations. The second volume of J. Boyle's two-volume set published in Aberdeen in 1777 opens with the poem about Newton, placing it before "Britannia." Throughout the century—not just in the years following Newton's death—the poem to Newton was an essential part of one of literature's most popular works.

The poem exemplifies the type of text written shortly after Newton's death to celebrate his scientific achievements and mourn his loss. Thomson's elegy for Newton begins almost exactly where Halley's ode ends, by posing a question about how Newton should be memorialized in writing. The poem thus sets up the dilemma of Newtonian popularization in works already aimed at a popular audience, such as the audience for Thomson's poetry. Thomson asks:

> Shall the great soul of Newton quit this earth,
> To mingle with his stars; and every muse,
> Astonish'd into silence, shun the weight
> Of honours due to his illustrious name?[64]

The witnesses to Newton's death are not the seventeenth-century Royal Society Fellows whom Halley portrayed in his ode, but an array of grieving women: the silent, shocked muses who find themselves unable to represent Newton's death appropriately. Thomson asks whether astronomy can fit into poetry at all through the use of a gendered frame where the passing of Newton renders female entities speechless. Thomson sets up the project of popularization as literally heavy, its weight reflecting Newton's published research. The poet aspires to the challenge of representing Newton because of his own desire to participate in "nature's general symphony," a symphony that requires sophisticated understanding of nature, mathematics, and music. With this symphony, Thomson aligns his work more with the aural origins of poetry than with print.[65] The opening also positions Newton as a welcomed guest in heaven, "mingling with his stars," a literal reference to Newton's advances in astronomy that evokes a heavenly register of sociability and community similar to Newton's divine position in the Halley ode.[66]

The timeline of Newtonian popularization in poetry deserves more critical attention than to automatically ascribe poetic inspiration to the *Opticks* at the exclusion of the *Principia*. Indeed, Thomson appears to have used Halley's ode as a source text, particularly those sections which focus on Newton's research in astronomy. This shows that the content of the *Principia* helped to determine what was in the poem, even if there is no document that literally tells us whether Thomson read it in full. One point in favor of its popularity in the 1720s is that Desaguliers cites Halley's ode in the footnotes to his poem on Newton, which was published in the same year as Thomson's.[67] One section of Thomson's poem is nearly identical to Halley's description of the moon. Where Halley had described the moon as "wandering Cynthia," Newton in Thomson's poem "fix[es] the wandering Queen of Night."[68] The description of the moon and tides that follows resembles a loose and embellished translation of Halley's ode, including the "yellow waste of idle sands" and an image comparing Newton to a sailor.[69] Thomson asks what "new wonders" Newton could possibly see in the afterlife following his many intellectual voyages. He describes the world Newton departs in terms nearly identical to those Halley used years earlier, writing of Newton that he

> While on this dim spot, where mortals toil
> Clouded in dust, from motion's simple laws,
> Could trace the secret hand of Providence,
> Wide-working through this universal frame.[70]

In Halley's words, "The things that so often vexed the minds of the ancient philosophers / And fruitlessly disturb the schools with noisy debate / We see right before our eyes, since mathematics drives away the cloud."[71] Halley wrote using similar metaphors of clouding and clarity, and he also compared Newton to other "mortals." Halley published more than a decade before the publication of Newton's *Opticks* (1704). The *Opticks* has been considered a landmark text in the popularization of Newton since the publication of Marjorie Hope Nicolson's *Newton Demands the Muse*, which provided a sustained focus on its links to poetry.[72] During Thomson's lifetime, the achievements of the *Principia* were still important enough that imaginative literature borrowed from Halley's poem introducing it.

As in Halley's ode, Thomson's poem makes important claims about Newton's masculinity as a touchstone for addressing scholarly communities. Thomson describes Newton's success in masculine terms: the ancient wisdom of the Greeks and Romans is compared to the "pride of boys" in comparison to Newton's achievements.[73] According to Thomson's metaphor, discoveries in astronomy can be explained in terms of individual development. Thomson reduces Euclid, Aristotle, and other ancients to children compared to the manly pride of Newton, claiming that the superior calculations in Newtonian science lent Newton himself maturity. When Newton's relation to the ancients is considered next to one of the dominant metaphors of both Halley's and Thomson's poems—that is, that mathematical success equates to superior masculinity—Newton becomes the man and the ancients become boys. Ben Jonson's poem memorializing Shakespeare uses similar metaphors, whereby Jonson cannot compare Shakespeare with Lyly, Kyd, or Marlowe, but instead ranks Shakespeare ahead of even "insolent Greece or haughty Rome," using terms describing immaturity as well as inferiority.[74] Lines like "Nature herself was proud of his designs" and "He was not of an age, but for all time" seem as much at home in flattering representations of Newton as they do in Jonson's poem to Shakespeare.

Similar flattering comparisons of Newton to other men occur elsewhere in this stanza, which uses masculinity to rank Newton opposite different communities of men. Thomson compares the knowledge of nature Newton has gained to the conquerors of empire, who usurp parcels of land "by violence unmanly."[75] Newton's work, by contrast, conquers nature through taming and subduing it. Much as the moon did in Halley's poem, Nature willingly surrenders herself to Newton's gaze as to a seducer: "Nature herself / Stood all subdu'd by him, and open laid / Her every latent glory to his view."[76] This is the kind of representation Sandra G. Harding described when she claimed that the *Principia* resembled a "rape manual," in which the universe is a frontier to be conquered—without its consent—by a dominant man in the form of Newton.[77] In these lines, Newton is able to claim what emperors and conquerors could not: an intimate physical knowledge of the entire world, intimate enough to call it his own. This is actually a contradiction in terms, because both intimacy and comprehensiveness would seem to deflate the Baconian inquiry the Society favored. One person could not possibly accomplish the combination of extensive and intensive knowledge of Baconian exploration, yet one book written by one man in one fixed location (Newton never left England) arguably had done so. This contradictory attribution comes from the normalization of extrapolation in the *Principia*, which had encompassed the great and small within its boards.

Fara writes about the placement of female entities in popular science: that "even though women were being coopted as audiences rather than practitioners of natural philosophy, the rational sciences were also portrayed as female muses. . . . Nature frequently appeared as a scantily clad woman whose hidden secrets could be unveiled by male experimenters."[78] This was certainly true in Thomson's poem and Halley's, so the coy seduction of nature does not seem to have changed substantially in the interim. The hyperbolic representation of nature revealing herself to the manly conqueror Newton fits with the unusual accolades Newton received during his life. Newton had been knighted on the basis of his intellect, his publications, and his work with the Mint, and he was in these capacities an eighteenth-century version of a conquering mythological hero.

Thomson's description of the system of the world dawning on Newton implies that this "all-piercing" penetrative knowledge came from a gradual process of illumination.

> All-piercing sage! who sat not down and dream'd
> Romantic schemes, defended by the din
> Of specious words, and tyranny of names;
> But, bidding his amazing mind attend,
> And with heroic patience years on years
> Deep-searching, saw at last the system dawn,
> And shine, of all his race, on him alone.[79]

The community compared to Newton in this poem is not founded on a collaborative, cooperative project of natural knowledge, but on gossip, noise, and chaos. Thomson explicitly criticizes the power of a community to relax standards of truth and method for its members and their "romantic schemes."[80] Whereas Halley had called attention to the priorities of organizations like the Royal Society, Thomson indicts community standards for truth directly. Halley had depicted Newton's discoveries in a manner consistent with individual experimentation—albeit on a divine level; Thomson questions the role of individual experiment as the foundation of scientific knowledge. After all, it is only after years of patient intellectual attention that Newton witnesses the system of the world. Patience, in addition to experimentation, fosters Newton's success. David Brewer's arguments about an eighteenth-century focus on community as opposed to partisanship suggest that Thomson's poem might tone down the appeal to partisanship present in the Halley poem. However, this is not the case here. If anything, the poem exaggerates the appeal to Newtonian partisanship, almost as if it is arguing for the transition of Newton to the heroic feats possible in fiction.[81] This is an interesting play in a poem that makes claims about masculinity; as with the example of the "unmanly" conquerors who try to take nature by force, Thomson aligns superior masculinity with qualities like patience and observation. This kind of masculinity would cast positive light on both natural philosophers and poets. This Newton not only conquers nature; his patience and years of observation make him also the most modest of witnesses.

Later, the poem imagines Newton's perspective and accompanies him on his "flight / Through the blue infinite" as he extrapolates the theory of universal gravitation to apply to the heavens and stars.[82] At the same time, Thomson makes claims about genius and community. Thomson identifies extrapolation as the source of Newton's genius: "thus to call / From a few causes such a scheme of things, / Effects so various, beautiful, and great, / An universe complete."[83] Newton's act of witnessing not only verifies his work; but in this passage Newton recognizes beauty, a source of the poet's inspiration and perhaps one of the reasons Thomson took on the project of writing the poem. Later, Thomson imagines Newton's inner work externalized, the natural philosopher himself out in orbit. Newton flies "first of men, with awful wing" around the universe.[84] Thomson speculates that what awaits Newton in the afterlife is the physical confirmation of the phenomena that he has already proved mathematically. "The heavens are all his own" because Newton knows and has defined them.[85] The poet even describes Newton as "our philosophic sun" at the center of inquiry regarding natural knowledge.[86] In Halley's poem, which uses the trope of the sun when it describes Newton as "Phoebus Apollo," the scientific community needs to cease its work in order to regard Newton's accomplishments.[87] In Thomson's poem, the community needs to pause and remember these accomplishments.

As much as Thomson's poem shares thematic unity with Halley's ode, Thomson frames the proof of universal gravitation differently apposite print culture than Halley had done. This was largely because the culture of print in Thomson's time was different from that of the *Principia*. In the interim, the lapse of the Licensing Act (1695) and the Statute of Anne (1710) meant heightened attention to print authority and copyright, as well as the centrality of the material text to these topics. Whereas Halley describes Newton writing the *Principia* as an act of witnessing and writing that supersedes Moses's transmission of the Ten Commandments, using words that suggest a culture of fast versus slow print, Thomson's Newton understands the laws of motion and universal gravitation due to patience and independent labor rather than witnessing. Thomson writes that Newton "from motion's simple laws, / Could trace the secret hand of providence, / Wide-working through this universal frame."[88] The Newton of these lines works with the laws of motion and observes phenomena, working almost as a scribe, "tracing" the "hand" rather than publishing a book. This is dissimilar to Halley, whose ode prefaced the poem's first edition and was not revised for subsequent editions. Thomson would have had the benefit of being able to read the General Scholium to the *Principia*, available after its second edition was published in 1713. The General Scholium gave more explanation of Newton's epistemological goals in book 3. Newton writes that "this most elegant system of the sun, planets, and comets could not have arisen without the design and dominion of an intelligent and powerful being"; lines like these can direct writers like Thomson to Newton's goals as a writer.[89] Newton describes his interest in science as witnessing a divine world-system. Thomson later joined his reputation to Newton's at the time he was completing *The Seasons* and beginning to establish his reputation. The materiality of the printed volume did not just unite celestial and terrestrial phenomena mathematically; it brought Newton's and Thomson's ideas together in one place, where at times they clashed.

One crucial difference between Halley and Thomson is that Thomson foregrounds his own reputation as the poet; in Halley's poem, Newton is described as a writer. If we are to look at this text as a significant event in the popularization of Newton, this is a critical decision. Now that readers are at the mercy of Thomson's interpretation, he is the reader's sole way to encounter Newton, filtered through form and mediated in print. This description invokes the poet's changing relation to print in the early eighteenth century. The primacy of Thomson-as-writer echoes in his verb choices: he imagines Newton singing, seeing, calling, naming, and dreaming—doing everything but writing. Thomson describes Newton as "All intellectual eye," building on Halley's ode, which had celebrated Newton as a witness, yet he also draws on the visual innovations of the later *Opticks*.[90] This unusual metaphor implies that vision is cognitive, that Newton's eye was somehow superior at thinking, and also places the poet in a separate space from Newton's sensorium. A poet's eye could still link to imagination, while Newton's eye was composed of intellect and the active thinking associated with witnessing.

Halley and Thomson had different perspectives on their print communities, for historical and personal reasons. Halley had published the *Principia* himself and therefore had a stake in the printed volume's success; Thomson, the son of a minister, was teaching and trying to network his way into the literary patronage system in the late 1720s. A breakthrough came with the publication of *The Seasons,* which Andrew Millar published in 1730 with the poem on Newton.[91] In 1729 Thomson made £105 and £137 10s for his sales of the four *Seasons* poems, the poem on Newton, a few other pieces, and the drama *Sophonisba*. This amount would have been enough to support him.[92] The receipt of sale Thomson sent to John Millan on July 18, 1729 acknowledged that the £105 included payment for "particular titles viz. Summer: A Poem, Autumn, a Poem, Winter a Poem, Britania (*sic*) a Poem, a Poem Sacred to the Memory of Sr. Isaack Newton; a Hymn on the Succession of the Seasons and an Essay on Descriptive Poetry."[93] The poem about Newton would have helped sales of *The Seasons,* because it was published when it was fashionable to write about Newton. The wealthy Halley lionized Newton as a figure in information culture; Thomson contemplated print as the medium he sought to disseminate his own art, and thus marked the medium for his own exploration while confining Newton to other modes of creative expression. Because of his different relationship to Newton and print, Thomson retains some of Halley's conventions while discarding others. One thread that remains is the link between Newton and an implied scientific community.

Thomson's poem identifies different types of homosocial community through which readers might contextualize Newton. One such community includes those who were Newton's peers in occupation but hardly his peers in knowledge, comparable to his colleagues in the Royal Society. Thomson writes that "but a few / Of the deep-studying race can stretch their minds / To what he knew."[94] Even within his intellectual community, Newton reaches only an elite group of thinkers whose minds are not sufficiently elastic to reach the heavens. This kind of definition is closer to popular stereotypes of Newtonian inaccessibility than the greater complexities of the *Principia* described in the previous chapter. The word "few" repeats a few lines later, this time referring to Newton's social circle. This other privileged community is "ye happy few / Who saw him in the softest lights of life."[95] Among these few who knew Newton intimately was John Conduitt,

Newton's nephew-in-law, known at that time to be at work on a biography of Newton. Thomson refers to the yet unpublished work as "the social passions smiling at thy heart," a reflective sociability characterized sympathetically in contrast to Thomson's earlier representation of "specious" scientific sociability.[96] Although the biography was never completed, Thomson humbly defers to Conduitt to translate Newton to a broader audience, casting himself as a poet distinct from Conduitt, the would-be popularizer.

Thomson does not describe himself as the best translator of Newton to an audience, but the popularity of *The Seasons* meant that this poem certainly reached a wider audience than anything written by Conduitt, and potentially a wider audience than other works of popular science. In the early eighteenth century, the print run of the well-known *Philosophical Transactions* was 1,200, whereas *The Seasons* sold many times that, and had print runs well into the thousands across the century. Sandro Jung's excellent book on *The Seasons*, print, and visual culture analyzes these different editions at length, with greater attention to *The Seasons* than to the poem about Newton. The poem would have made a good introduction to Conduitt's biography of Newton, something that would have enhanced Thomson's reputation and bolstered sales of *The Seasons* further. It certainly resembles a prefatory poem in style, with its elevated subject and its concise length. The deference Thomson shows to Conduitt in this passage further implies that prose was treated differently by eighteenth-century writers than poetry was. As prose fiction was beginning to gain a stronghold in the print market, Thomson's poem takes time to pay its respect to prose and its own "social passions."

Even though much of Thomson's poem describes Newton as a solitary and unique figure, Thomson does not allow Newton to wander the afterlife alone; instead, a mythical community is conjured to incorporate Newton, a community most likely composed of other men. The women in this poem are either the unruly celestial bodies who must be tamed, a saucy Mother Nature who lays herself open to Newton, or the Muses whose inability to memorialize Newton opens the poem. By the poem's end, Newton becomes a member of a community of unnamed geniuses like him, and he has found a place of mutual understanding. "Now he wanders through those endless worlds / He here so well described, and wondering talks / And hymns their Author with his glad compeers."[97] The explicitly masculine elements of the poem have moderated by this point; even so, these lines suggest a masculine community where Newton and his implicitly male "compeers" gather to share their ideas in the afterlife. Newton is as happy and sociable in the heavens as on earth, with the universality of phenomena he described in the *Principia* applied to his own circumstances.

Scientific Communities and Homosocial Science

Within the context of scientific poetry, the poems by Halley and Thomson distinguish themselves while still remaining connected thematically to works about other scientific figures, especially when it comes to the need to represent a homosocial scientific community. Samuel Butler's poem "The Elephant in the Moon" (date unknown, est. 1666–80) satirized the activities of the Royal Society, defined as a plural subject, a homogenous community. The subject of the poem, "A learned society of late," attempts to survey the moon by telescope, "For settling of new plantations, / If the society should incline / T'attempt so glorious a design."[98] Here, too, the understanding of celestial phenomena is couched in masculine terms—those of settler colonialism. Butler uses metaphors similar to Halley's when discussing the Royal Society, setting up the same kind of contrast between masculinity, science, and community that Newton would later be said to rise above in Halley's ode. Butler would later revise this poem to change its meter from iambic tetrameter to iambic pentameter. Both versions as well as an additional fragment are published in *The Poetical Works of Samuel Butler.*[99] My examples are drawn from the earlier tetrameter version.

The section where the men line up to look at the moon evokes a masculine community together vying for domination of a feminized body.

> When at the full her radiant light
> And influence too were at their height.
> And now the lofty tube, the scale
> With which they heaven itself assail,
> Was mounted full against the Moon;
> And all stood ready to fall on,
> Impatient who should have the honour
> To *plant* an ensign first upon her.
> When one, who for his deep belief
> Was virtuoso then in chief,[100]

The sexualized imagery ("mounted full against the moon," "lofty tube," "ready to fall on") and the jokey sexualized rhyming of "honour" and "upon her" show that there were masculine stakes at play in the ostensibly gentlemanly environment of the Royal Society. What was seen as manly science within the Society was not universally admired from without. The jarring representation of lecherous would-be virtuosi intruding into the territory of poetry—especially the moon, which represents both a poetic and scientific subject—is comic and grotesque at the same time. Instead of pursuing a female rival, the philosophers attempt to seduce the moon, which makes them seem even tinier than the mortals they already are. A "virtuoso then in chief"—implying that the term is interchangeable and unstable—penetrates the moon with his gaze through the union of eye and telescope. Streater's press also yielded the dildo-making virtuoso of Rochester's *Farce of Sodom,* so this association of virtuosi with confused sexuality is not unique to poetry. Shadwell's *The Virtuoso* also addresses the theme, when the supposed virtuoso Gimcrack fails to notice that he is being cuckolded. Butler's philosophers, for their part, are much like those criticized in Halley's poem as being part of an inchoate mob. It is not difficult to imagine these characters, who are so busy interacting with one another—in debate or experiment—that they lose sight of foundational philosophies.

Later works about Newton on the Continent present different perspectives on gender and community, so Thomson and Halley—and to some extent Butler—can be set apart from Continental representations of Newton. A French Newtonian poem, Louis Davy de la Fautrière's (1700–1755) "Épître Newtonienne" (1739), surfaces in J. B. Shank's discussion of anti-Newtonian works, so Davy de la Fautrière's relationship to Newtonianism as a project of popularization is less evident than in Britain.[101] We know that Davy de la Fautrière was a Jansenist magistrate or sympathizer "as of September 1754" and died shortly thereafter.[102] A catalogue of his books was published in 1756.[103] In contrast to the British poems, community has less of a stake in this poem, which focuses strongly on attraction. Davy de la Fautrière subtitled the poem "Sur le genre de philosophie propre a rendre heureux," or "On the genre of philosophy most suitable to make one happy." The poem is rumored to be connected with Madame de Graffigny, who knew Davy de la Fautrière; regardless of who the female dedicatee is, she is addressed as "Uranie," muse of astronomy and daughter of Zeus. Davy de la Fautrière refers to the "genie / De l'Hôpital, et de Newton," so the poem dialogues British science with the works of French thinkers like de l'Hôpital.[104] "Épître Newtonienne" contains some significant passages that relate to Newton—different from others found in Newtonian poetry.

A passage from early in the poem includes an interaction between the speaker and the voice of Newton himself, who speaks from the afterlife about universal gravitation:

> J'interroge Newton, et telle est la réponse:
> «Sur un centre donné que tout pese à la fois,
> «D'une attraction mutuelle
> «Que tout corps éprouve les loix,
> «Voilà la cause universelle;
> «Nul mobile dans l'Univers
> «N'échape à ces loix générales,
> «Du combat des forces centrales
> «Naissent les mouvements divers.
> Content d'une leçon si claire,
> Mon esprit goûtoit la douceur
> De comprendre enfin ce mystere
> Des ouvrages du Créateur;[105]

The attractions of a Newtonian universe that "Porta mon esprit et mes yeux," transporting the speaker's spirit and eyesight, invoke an address to Newton himself, whose spirit responds in a brief poetic lecture on his own theories.[106] In this poem, curiosity about the Newtonian universe distills to the attractions between bodies, the subject of the *Principia,* rather than to Newtonian optics. "D'une attraction mutuelle / Que tout corps éprouve les loix, / Voilà la cause universelle." All bodies prove these laws of mutual attraction. In a poem addressed to a woman, the attraction of bodies has both seductive and explanatory overtones. Attractive bodies would thus include the body of the woman to whom the poem was addressed under the guise of Urania. Newton's voice, cast in this manner, becomes an instrument of seduction as well as explanation, and the natural philosopher provides his own poetic popularization of his theories. The same topics would be treated differently in contemporary works like Algarotti's, discussed in the next chapter, where the would-be seducer or popularizer replaces Newton's voice with his own voice and body. In "Épître Newtonienne," the speaker, satisfied with Newton's simple response, nonetheless proceeds to wonder exactly why the planets move as they do.[107] His intercourse with Newton was sufficient even if it did not sufficiently teach him about the Newtonian universe. Later in the century, a different French writer compared Newton to Descartes in terms that showcased the former's careful attention to detail. In 1765 Antoine-Léonard Thomas published a prose elegy to Descartes, writing that "Newton donnoit aux plus petits details l'empreinte du genie"—"Newton gave the imprint of genius to the smallest details."[108] The only other use of "empreinte" in Thomas's text also engages Newton as one of many writers Descartes's "imprint" influenced, so the term "imprint" is linked with Newton in its only other usage.[109]

This chapter has largely centered on masculinity, any discussion of which necessarily has implications for the treatment of women. We have seen that women are not given autonomy in most poetry that mentions scientific topics, but one exception to this appears in a poem that mentions Newton briefly but significantly. Newton appears in Christopher Smart's "The Temple of Dulness" (1745) in a stanza about the goddess Mathesis, who seems far less eroticized than the feminine entities in other poetry that discusses Newton. As a point of comparison, Pope's Mathesis in the *Dunciad* (1728), Smart's influence, had been depicted as follows:

> Mad Mathesis alone was unconfined
> Too mad for mere material chains to bind
> Now to pure Space lifts her ecstatic stare
> Now running round the Circle finds it square.[110]

Mathesis's mobility resembles some of Eloisa's mobility in Pope's "Eloisa to Abelard," whose "loose soul [springs] all unbounded" to her absent love while she dreams.[111] Mathesis, for Pope, is inseparable from her namesake: abstract, immaterial, and geometric, she may be mad because she is not fully embodied and therefore cannot fully have a mind.

Smart's Mathesis differs from Pope's, and her relationship to Newton is significant in light of Newton's comparison to feminized entities in other poetry.

> This Mathesis her feet all bare,
> Ungirt, untrim'd, with dissoluted hair:
> No foreign object can her thoughts disjoint;
> Reclin'd she sits, and ponders o'er a point.[112]

This Mathesis approaches a neoclassical representation of Newton, calling to mind the Godfrey Kneller portrait of Newton when he was first made an MP, wearing his robes, bareheaded, hair uncoiffed, devoid of ornament. The description of Mathesis's focused attention span echoes anecdotes told about Newton, including a practical joke where a friend cleared Newton's plate before he had eaten, and he allegedly looked down at his empty plate, thinking he had eaten dinner but had forgotten about it. Likewise, Mathesis cannot be swayed by the objects of sense perception.

However, later in the poem, Mathesis is forcibly displaced outside of scientific innovation:

> She, when great Newton sought his kindred skies,
> Sprung high in air, and strove with him to rise,
> In vain—the mathematic mob restrains
> Her flight, indignant, and on earth detains;
> E'er since the captive wretch her brains employs
> On trifling trinkets, and on gewgaw toys.[113]

In this passage, Newton endeavors to rise to the heavens, but the goddess's attempt to accompany him is thwarted by other men in the scientific community, the same male "mob" we have seen repeatedly in other guises throughout this book. Earlier in this stanza Mathesis is described as a "monster, arrogant and vain" for her attempts to use "strange diagrams[,] . . . right lines and curves, with figures straight and round" to teach mathematical principles.[114] The negative depiction of the scientific community in these lines doubles as a representation of women's blocked access to natural philosophy. Her wings clipped, a prisoner on earth, Mathesis must play with stereotypically feminine articles, with objects rather than ideas, her brain imprisoned by its fixation on baubles. Her mind cannot be free and unbound as Pope's Eloisa's can, nor can she be a reliable interpreter of visual representations of abstract ideas. The mathematical diagrams she tries to interpret belong to a male community who also restricts their interpretation. Newton, of course, experiences his apotheosis in Halley's, Thomson's, and Smart's poems. Smart's poem, perhaps more than others, shows that these representations of Newton as a hypermasculine figure do not come without cost; rather, they exemplify the masculinization of science, and the channeling of women's interest in science—and abstract thought itself—into insignificant pursuits.

Thomson's poem had built on Halley's ode in its exploration of Newton's gender and his relationship to a community during a time when Thomson's relationship to the print marketplace, unlike Halley's, was as a little-known poet. Thomson's poetry therefore suppressed print as a factor in Newton's public persona, even though the publishers of his books celebrated Newton—sometimes as much as they celebrated Thomson. The poems articulate their position on masculinity through comparing Newton to other members of his community and by defining Newton's masculinity relative to print history. Thomson's Newton ultimately takes his place among the planets and stars, where Halley had initially directed readers to look for him. Both works show a Newton who occupies a position somewhere in between fellow mortals and God, marking Newton's superiority over his contemporaries. The poem implies that the scientist and poet may yet connect over their (gendered) appreciation of divine perfection. Following this logic, perhaps it is Apollo, who watched over poetry as well as truth and light, and whom Halley also invokes as Phoebus Apollo in his ode, who best captures the kind of divine, masculine persona Thomson imagines for a scientist who had such far-reaching influence.

Men of Poetry, Men of Science

More than two hundred years after Newton's death and Thomson's elegy, Allen Ginsberg wrote the *Howl* poems, elevated as foundational literary achievements of the Beat Generation. The Beats were iconoclastic in content and accessible in language; they participated in experimental performances where live readings were places of witnessed and valuated experimentation. These experiments in language—rather than science—echo the earlier witnessed experimentation of the Royal Society, which also took place in locations that were semi-public and semi-private. Anne Hartman traces the roots of Ginsberg's confessional poetry to the eighteenth century, centering on figures like Rousseau and Kant and their engagement of publicity, privacy, the impersonal, and the personal.[115] Even earlier in the eighteenth century, these tensions inflected poetry about Newton.

The poetry about Newton I have described in this chapter shares a focus on publicity and gender and how to treat the personal and the professional. It follows that such poetry should resist confessional depictions because a third party, the poet, represents the would-be confessor, Newton. Even so, the question of what Newton could or might confess—latent confessional content—persists in representations of him.[116] Wordsworth's depiction of the statue of Newton at Trinity College, "Voyaging through strange seas of thought alone," uses the mystery of Newton's thought as a cognate for the inscrutability of the "marble index" of his expression in statue form.[117] In *Howl*, Ginsberg portrays a figure who is "obsessed" with "alchemy [and] the use of the ellipse" as well as "catalog" and "meter," running around in the freezing weather, oblivious to his surroundings because of his obsession.[118] Hartman analyzes Ginsberg's "use of the ellipse" as a metonym for confession, as a generalized "outing" of controversial content. Ginsberg's ellipses transcend their definition, because to articulate an ellipse is to give weight to the textual equivalent of a vacuum: "an ellipsis is a figure implying what has been omitted, and carries added significance in a poem with a confessional premise of disclosure."[119] In that capacity, the weight of omission in Ginsberg's poem parallels the emphasis on what *might* be confessed in poetry about Newton, a potentiality that is ultimately replaced by a communal or artistic interpretation. Ginsberg's reference to disclosure comments further on space and time: the speaker of the poem is "not being safe and running through the streets, obsessed with the ellipses. These ellipses are not merely gaps, but 'catalog' rhythm in time and capture the space of the plane; although they are an absence, they have effect."[120] The presence of absence and the weight of space resemble poetic attempts to represent the unrepresentable in Newtonian science and recall my discussion of George Horne's passage about Newton in the introduction, where the fetishism of abstraction is at odds with literal representation. Ginsberg's typographical ellipse resembles the elliptical orbits Newton worked with mathematically.

Howl, with its synthesis of science, religion, and poetic forms, calls to mind representations of Newton's genius that surfaced at the time of the *Principia*'s publication and during the earlier part of the eighteenth century. Ginsberg's lines seem like chiastic counterparts to popular Newtonianism: if "Newton demands the Muse," so, too, can the composition of poetry resemble scientific innovation.[121] Ginsberg describes a figure who "dreamt and made incarnate gaps in Time & Space," innovative practices that scientists and poets share.[122] These lines correspond to Newton's rumored monomania about scientific research: an ingenious synthesis of mathematical, theological, and alchemical work that may have contributed to his nervous breakdown in the 1690s. Ginsberg describes flashes of genius in linguistic and poetic forms, rendering time and space as text, and linking inspiration with the "Pater Omnipotens Aeterna Deus"—invoked in Latin, like the God of the *Principia*.[123] Loni Reynolds describes *Howl* in the context of cognitive and disability studies, in which the representation of cognitive disability frames "an alternative mindset that provides creative insight and poetic inspiration."[124] These words resonate in my discussion of Newton and poetry, where published works about Newton proffer descriptions of genius as an alternative mindset resembling those in the lines above. In this textual synthesis, innovation is a form of prophetic knowledge, a trope that links Ginsberg's conception of genius to those written by Newton's contemporaries. Newton, too, was able to connect forms and seemingly unrepresentable images, making incarnate his own juxtapositions of time and space in the *Principia*.

The following chapter will address the popularization of Newton in English-language editions of texts that address a gendered audience. We are, to some degree, leaving Newton himself in the past at this point in the book and focusing on his cultural translators as popularizations emerged later in the eighteenth century.

4 Extending Popularization in English-Language Editions of Francesco Algarotti's Works

This chapter studies the tension between the production and transmission of Newtonian ideas by evaluating the editorial choices and diffuse scientific content in works by Francesco Algarotti that circulated in popular editions in Britain during the mid-eighteenth century, including *Il Newtonianismo per le dame* (1737; *Newtonianism for Ladies,* first English edition, 1739). Algarotti not only explained Newton to the ladies of Europe: he also wrote similar texts demystifying painting and opera.[1] His works helped to elucidate these sister arts for an emerging middle class with a desire for sophisticated cultural knowledge.[2] In spite of the seeming straightforwardness of these popular texts, there was a gap between the stated aims of Algarotti's printed works and the ideas they actually conveyed, a gap that scholars have not explored.[3] Algarotti interwove scientific information with cultural and aesthetic criticism, incorporating concepts such as attraction, desire, and extrapolation seamlessly.[4] The kind of reading public—gendered, sophisticated, and even sexy—Algarotti conjures in *Il Newtonianismo per le dame* failed to predict the composition of the readership English publishers imagined when they printed different translations and editions of his works.[5] Algarotti's influence on the reception history of science in England is more widespread and diffuse than has been assumed.[6] My focus on editions of this Newtonian "popularization," as well as non-scientific works by the same author, links translation, editorial practices, and transmission to evaluate the print context of popular texts that *inexpert* English-language readers encountered across the century.

Not long after the *Principia*'s publication in 1687, Bernard Le Bovier de Fontenelle's *Entretiens sur la pluralité des mondes* (1686), a popular science bestseller, circulated in London with a new English translation by playwright Aphra Behn, published as *Conversations on the Plurality of Worlds* (1688). While Newton had positioned himself as a uniquely gifted mathematician with a divine access to the truths of the universe in the *Principia*, *Conversations* reveals that a secondary marketplace was developing for explanations of scientific ideas that catered to nonspecialist readers in English. These translated works—like Fontenelle's metaphorical translation of Cartesian principles into easy and pleasant conversations—found a stronghold in the print marketplace and rose in popularity following Newton's death in 1727. These books include Algarotti's *Il Newtonianismo per le dame*, a text that is commonly thought to have helped "popularize" scientific ideas in Britain and on the Continent by explaining complex Newtonian natural philosophical concepts to a nonspecialist, literate audience without the use of mathematics. However, as scholars of print know, publication does not equal dissemination; the existence of a printed book does not guarantee the spread of its ideas. Roger Chartier writes that editorial conventions attempt to create and anticipate "specific reading attitudes."[7] Specifically, the publication of a work that explains a sophisticated scientific concept—a "popularization" of science—implies that we should consider its associated printing practices and reception closely.[8] Math was frequently figured as masculine in Newtonian popularization, as we saw in the poems from the previous chapter. In *Newtonianism for Ladies*, a lack of mathematical content facilitated the text's mobility across genders and audiences.

Most critics have written about *Il Newtonianismo per le dame* as a single text that translated Newtonian science to a gendered reading public. Yet the differences in the text's eighteenth-century English-language editions (1739, 1742, 1765, 1772) are crucial to understanding its complex transmission across the mid- and late eighteenth century. These editions reflect publishers' responses to imaginative literature's influence on popular tastes and the changing role of the dialogue in education.[9] As in other aspects of publishing, where "growth was particularly driven by the increasing population of London and rising literacy rates, which heightened demand for print," this heightened demand corresponded to the production of diverse editions whose publishers seemed completely unaware of the editions' similarities to one another.[10] By evaluating the textual variations across editions of *Newtonianism for Ladies,* I show how its transmission blurred distinctions between popular science and popular literature, during a time when audiences for both overlapped. I then explore the connections between science and the sister arts in Algarotti's other works published in English, which disseminate popular science even further afield by packaging it in different cultural contexts: those of art connoisseurship and music.

English-Language Readers of *Newtonianism for Ladies*

Some contemporary scholars of eighteenth-century literature, print, and the history of reading have paid increasing attention to reading publics in their research, whereas others have examined the paratexts that inform the construction and reception of eighteenth-century texts.[11] Still other critics have quantified evidence to contextualize readership in terms of actual publication, circulation, and sales.[12] Works that focus on the book history of science, by contrast, investigate the circumstances surrounding the production of scientific texts more frequently than they study the readers of those texts.[13] Even contemporary scholars who do focus on readers of scientific texts focus on ideal readers or the histories of the books themselves, rather than on the links between books and the reading publics they have implicitly or explicitly addressed. These texts were, to use Roger Chartier's phrase, "unequally suited to reappropriations" because of their diverse audiences and the different editorial choices that shaped them.[14]

Why were so many distillations of Newton's ideas popular with diverse audiences? The previous chapters have shown that Newton's work in the *Principia,* while superficially coherent to a nonspecialist reader who knew Latin and Euclidean geometry, was nonetheless inaccessible to many English readers. Regardless of this barrier to access, the English still had opinions regarding Newton. The debate between Cartesian and Newtonian natural philosophy early in the eighteenth century among inexpert English readers suggested nationalistic assumptions rather than scholarship, as Voltaire reveals: "Very few people in England read Descartes, whose works indeed are now useless. On the other side, but a small number peruse those of Sir Isaac, because to do this the student must be deeply skilled in the mathematics. . . . Notwithstanding this, these great men are the subject of everyone's discourse. Sir Isaac Newton is allowed every advantage, whilst Descartes is not indulged a single one. . . . Newton is here as the Hercules of fabulous story, to whom the ignorant ascribed all the feats of ancient heroes."[15] Most would agree that Newton's reputation for his intellectual labors approached a Herculean status in eighteenth-century popular culture; however, Newton's writings themselves were not simply difficult in content. His work, especially in the *Principia*, presented a unique challenge to popularization. The difficulty of Newton's mathematical calculations meant that even though his work was hard to explain fully, his popularity and intellectual significance necessitated public translation. Fara writes that "debates about religion and philosophy were saturated with national interests, and patriots who knew little about optics or gravity fêted Newton because he was an Englishman."[16] Few literate people were so "deeply skilled in the mathematics" as to fully comprehend the *Principia;* even so, the English had enough to say about Newton and René Descartes that the two remained central subjects of discussion. The idea that Newton could be "allowed every advantage" against Descartes, without debate or argument, derives from Newton's status as someone "to whom the ignorant ascribe all the feats of ancient heroes," and who evokes nationalist pride. Paradoxically, a nation that took pride in its scientific achievements championed Newton's achievements without understanding them. In order to rectify this imbalance, people read works like Algarotti's Newtonian popularizations, which—to borrow the Herculean metaphor—compose a many-headed Hydra on their own.

The absence of expertise that Voltaire described explains why popularizations of science found a niche in the English print marketplace; Newtonian natural philosophy was narrated to the English through a diverse array of publishing practices as well as through the published works themselves. *Newtonianism for Ladies* addressed an imagined diverse audience for eighteenth-century natural philosophy by connecting Newtonianism to pleasurable entertainment and by explicitly engaging an audience of women.[17] Much as the English frequently discussed the works and ideas of Newton and Descartes without reading them, scholars have written about the English editions of Algarotti's *Il Newtonianismo per le dame* without comparing them, treating them as interchangeable, or even identical, editions. Sarah Hutton analyzes the variations in Italian editions from the 1730s through the 1750s, working under the assumption that later English editions were identical to Elizabeth Carter's first translation from 1739.[18] Mirella Agorni describes these editions as "reprints" of Carter's translation; Robin Valenza acknowledges that titles were altered but mentions no other differences.[19]

As the following chapter shows, *Newtonianism for Ladies* was not a single English text, but several unique editions. In diverse ways, these editions manipulated form and content to orient popular Newtonianism to different, gendered reading communities.

The differences between eighteenth-century English-language editions of *Newtonianism for Ladies* reflect publishers' responses to different generations of female and male readers, as well as distinct shifts in the landscape of eighteenth-century British popular science, far more than has been acknowledged.[20] In contrast to Hutton's explanation that publishers designed variations in Algarotti's Italian editions "to disguise the fact that [Algarotti's] elegant phrases contained unpalatable opinions about the nature of the universe," my examination surveys a different market in England because of the wide acceptance of a Newtonian perspective.[21] These supposedly "unpalatable" features concerned the British far less. Paula Findlen's important work on the ways Algarotti exemplified a form of "authorship without authority" also influences my argument here, specifically as I explore connections between this displacement of authority and the gendering of Newton in eighteenth-century British culture.[22] This "authorship without authority" tacitly legitimated the variable treatment of Algarotti's works in printed translations. Indeed, when we consider the centrality of Newton as author—in the *Principia* and in Halley's poem—the extensive variations among editions reflect important connections between literature and science in the eighteenth century. This relationship facilitated the unique mixture of "coquetry mingl[ing] with science," as Valenza puts it, which characterizes Algarotti's text.[23] The marketing of these printed works is linked to the broader reception history of Algarotti in England and extends to other works by Algarotti, including his essays on opera and painting.

Algarotti posits a different kind of gendered community of print from the masculine print communities Halley and Thomson recognized and the binaristic depictions of gender and science in Samuel Butler and Louis Davy de la Fautrière.[24] *Newtonianism for Ladies* triangulates the relationship between the masculine expert in the form of Newton, the audience—composed ostensibly of women, but really reflective of a broader polite community—and the translator of Newtonian expertise, the narrator, representing Algarotti's narrative persona. With this more decentered way of orienting published explanations of Newtonian science, mediated by a translator, to an audience, earlier binaristic thinking about a hypermasculine scientist like Newton conquering Mother Nature no longer applies. If anything, the feminine aspects of nature and natural philosophy are foregrounded and manipulated for readers' easier immersion in the challenging topics of Newtonian science and optics.

The editions I study include the 1739 first English translation of Algarotti's text by Elizabeth Carter, widely marketed by *Gentleman's Magazine* publisher Edward Cave. The 1742 edition George Hawkins published combines a piracy of Carter's translation with an unattributed—and previously unidentified—translation of the first French translation by Louis-Adrien Duperron de Castera from 1738. Later in this chapter, I use Robert Urie's 1765 Glasgow edition, which downplayed the gendered aspects of Algarotti's text and audience, even though this book, too, used Carter's more or less literal translation of Algarotti. I also analyze Francis Newbery's 1772 edition, which, in spite of some of its racy dialogues, was sold alongside the children's science book *Tom Telescope* (1761). The diversity of these English-language editions of *Newtonianism for Ladies* challenges scholarly conceptions about what eighteenth-century audiences knew from reading a single popularization and reveals how English-language publishers imagined different markets for popular science that had varying degrees of overlap with imaginative literature.

Algarotti's Continental reception diverges from his reception in Britain, Newton's home nation, where teaching texts helped the British to understand one of their own cultural heroes. Algarotti was a natural philosopher and art critic who was a friend of Voltaire, a Fellow of the English Royal Society, and a patron of the arts. His popularization of Newton's work on optics, as modern scholar Massimo Mazzotti writes, aided in "Algarotti's self-fashioning as an independent cosmopolitan author[,] . . . as a skilled experimenter and one of the champions of Newtonianism on the continent."[25] *Il Newtonianismo per le dame* was read by Enlightenment Europeans as a book spotlighting Algarotti, perhaps as much as it showcased Newton. But as much as this work aided the development of *Algarotti*'s public persona on the Continent, the text was popular in England because England is where Newton had lived, published, and died and where he was widely acclaimed as the nation's greatest intellectual giant. There, Algarotti was essentially a handmaiden to the clarifying of Newton's genius for a popular audience. *Il Newtonianismo per le dame* could appeal to both men and women because, according to J. B. Shank, Algarotti "positioned himself as a bridge between serious male learning and the feminized world of *le beau monde*."[26] Valenza follows this argument in what she sees as a developing gap between expertise and readership in Newtonian texts of the eighteenth century. The relationship between gender, masculinity, and community was therefore different in Algarotti's text than in works I discussed in previous chapters.[27] As a teaching text, *Newtonianism for Ladies* exhibits the shift Michèle Cohen describes in eighteenth-century didactic literature.[28] Using the example of French-language textbooks marketed to the English, Cohen identifies the trend away from conversation and toward grammatical instruction in the teaching of French. The editions I analyze demonstrate similar trends. They begin as conversations, and the editions proceed in two different directions throughout the eighteenth century: the novelistic and the didactic. These editions found different ways to address evolving reading publics, and they contextualize the evolution of both pleasurable and instructive literature.[29]

Pleasurable Entertainment and Popular Science

There was, then, a lacuna of works for a gendered audience explaining Newton's works in entertaining and even gripping ways. That there could be such a subfield in publishing suggests how much an eighteenth-century popular audience wished to grasp Newtonian concepts, and how eager publishers were to meet the demands of this market. Hutton describes Algarotti's text as "evidence of a female readership for scientific writings."[30] Algarotti used a series of conversations between a marchioness and her tutor to outline Newton's achievements in optics and physics, avoiding the use of math and employing comparisons from popular culture in order to provide commonplace references for physical or abstract phenomena. As Franco Moretti writes, definite articles in titles "announce a noun as something we already know," as with "le dame," or "the ladies" in this instance; by contrast, "the indefinite suggests the opposite," that a topic is new or unusual.[31] The title of Algarotti's text, *Il Newtonianismo per le dame*, implies that the author understands women, and that the instruction is one-sided. In short, Algarotti not only knew Newtonianism, he also knew "the ladies" as an audience.[32] His work did have a broader audience beyond women. Inexpert audiences, male and female, who were interested in Newtonian science but who lacked mathematical expertise could gain general knowledge from reading it.[33]

My treatment of Algarotti's text differs from that of prior writers. Literary scholar Mirella Agorni interprets the title's address aimed at the "ladies" as a proxy in which "women were used as a kind of rhetorical device to represent the cultural exclusion of the ideal readers."[34] Algarotti's text was not, however, responding to widespread cultural exclusion because he focused on Newton's *Opticks,* a more accessible text than the mathematically complex Latin *Principia* (1687). The *Opticks* had been popular in English, Latin, and Italian in several editions and its conceptual transparency (relative to the *Principia*) did not exclude many readers. Beyond cultural exclusion, Algarotti's work responded to the tastes of a particular reading class for entertainment reading rather than didactic reading. The audience of *Il Newtonianismo per le dame* could have read and understood much of the *Opticks.*[35] As women were a significant audience for books purchased as pleasurable entertainment, it is not surprising that publishers took advantage of an opportunity to market to them.

Because it is unfamiliar to most readers, I will examine Algarotti's text briefly before analyzing the differences and commonalities between its several English-language translations and adaptations.[36] The introduction to *Il Newtonianismo per le dame* orients the figure of Newton, as well as Algarotti's own content, toward an imagined audience. He explains the book's entertainment value and compares it to a different spectacle from scientific experiment, one more feminine and entertaining. "I have endeavored as much as possible to render it lively, and make my readers interest themselves in it as they would in a composition for the Theatre. Is there any thing, especially where ladies are concerned, in which a writer should omit any endeavors to move the heart?"[37] Algarotti compares the immersion possible for an audience of live theater to the immersion in narrative his readers will feel, implying that learning science in another way is dull and lifeless. As we have seen earlier in this book, theater was a varied trope for representing the representability of scientific content; in Horne's satirical description of abstract theater, drama and science do not align well. Here, using dramatic structure to describe scientific content is more successful. These theatrical comparisons preview the book's structure. It is organized in dialogues of ascending complexity and sophistication, much like a complex plot, interspersed with flirtations between the narrator and the marquise. For the narrator and the marquise, a mutual interest in science replaces heterosexual attraction, his tutelage takes the place of a courtship narrative, and the marquise's happy understanding of Newtonian science concludes the narrative like the end of a marriage plot. Franco Arato claims that Algarotti's work helped to create an audience for this mixture of science and entertainment.[38] As Hutton captures it, people wanted "Newton without numbers."[39] Like early novels that granted their audiences pleasure and instruction together, *Newtonianism for Ladies* promised similar satisfaction, but instead of receiving instruction under the guise of pleasure, readers were granted pleasure labeled as instruction.

Its introduction defines the readership of *Il Newtonianismo per le dame* in mixed classical and religious terms, establishing the narrator's authority and displacing Newton. Algarotti writes that "the sanctuary of the temple will always be reserved for the priests and favorites of the deity; but the entrance and its other less retired parts will be open to the profane."[40] These "priests and favourites of the deity" included some who worked directly with Newton—including Conduitt, his nephew-in-law.[41] Whereas in prior literary accounts Newton had been treated as a mortal with access to divine wisdom, Algarotti's Newton is himself the deity, his knowledge and expertise far above that of the women Algarotti addresses.[42] Provocatively, the spaces that Algarotti describes as masculine are the most private and interior spaces of this temple, apposite "profane" spaces where women are permitted. Elevating the presentation of Newton allows the text's male narrator to retain his position as an authoritative teacher in addition to his role as the romantic interest of the story. The more public spaces of Algarotti's imagined temple are not simply sexist, but also define his readers' limitations: as tourists, they linger at the entrance of the temple together, experiencing intimacy with Newton through the narrator's mediation.

Patricia Fara calls *Newtonianism for Ladies* "greatly simplified" and "less ambitious" than other works of popular Newtonianism because it could not lend itself to research and offered only "superficial knowledge" to its readers.[43] These sentiments align with those of other historians of science, and partly explain why this book is overlooked. Valenza describes *Newtonianism for Ladies* as the place where "science comes closest to *belles lettres*."[44] I am inclined to agree with Valenza insofar as *Newtonianism for Ladies* fulfilled a different ambition than Fara allows: its ambition was to make Newtonianism pleasurable and entertaining, and it succeeded.

Algarotti's popularization capitalizes successfully on pleasurable modes of discourse, such as those of small talk, courtship, and seduction, concepts also central to imaginative literature. By linking these attributes with concepts like attraction, Algarotti fuses scientific popularization to the kind of erotic tension common to imaginative literature. An example from dialogue 2 epitomizes the discourse of two people pretending to talk about a socially acceptable topic—in this case, optics—while trying to flirt with one another. The conversations between the narrator and the marquise often digress from scientific explanation in order to wax flirtatious. In one example, women's "bedroom eyes" are explained scientifically as follows: "in the dark, the pupil is more open and dilated, which makes the eyes look blacker and brighter in the night than in the day when the pupil is more contracted. How many eyes have triumphed in the evening and gained conquests which they lost the very next morning at the sight of the sun!"[45] In spite of these attempts to connect Newtonian science to women's social experiences, the author still reduces those experiences to a series of balls, parties, and flirtations. With the repetition of examples that relate to seduction and courtship, Algarotti's work, ostensibly meant to popularize Newton, could even serve as a seduction manual, where Newton is the chaperone who legitimates covert flirting, rather than the gendered, hypermasculine figure we have seen previously.[46]

Throughout the text, the connections the narrator makes among Newtonian ideas parallel the narrative's attempts to unite the marquise and the narrator. After her initial resistance to the narrator's ideas, the marquise praises his ability to modulate his "philosophical enthusiasm, which had carried him so far," in a register that mirrors courtship discourse.[47] As we saw in the publication history of the *Principia*, Newton's work on action at a distance distanced the writer himself from his readership. That manipulation of distance comes into play in *Newtonianism for Ladies* through the narrative's small-scale flirtations and the broader narrative of the narrator and marquise, who can never close the distance separating them, however attracted they are to one another's ideas. The narrator plays with the idea of action at a distance as he makes a point about variations in perception between the two eyes of one person, claiming that "we will bring them gradually nearer to us, till at last we will agree to believe some difference between your ideas and mine, and from thence between the two eyes of some persons to whom the same object appears bigger when seen through one eye, than it does when looked upon by the other."[48] Algarotti's text interrogates the attraction between two distant people, finding tension between a man and woman whose ideas may differ, but who are set beside one another. The push and pull in this interplay of differing perspectives evokes the physicality of both the narrator and the marchioness: the two of them contemplate these phenomena in the same intimate space. Her response could also be a model for flirtatious behavior that refrains from crossing lines of propriety. "How is this possible, said the marquise? There is no end to your visionary fancies, and you seem resolved to put me to the utmost proof of my credulity . . . I must confess I think this a very daring way of proceeding."[49] Although she thinks the narrator is daring in his proceedings, the marchioness follows him, herself attracted to these "visionary fancies." The intellectual capital of Newtonianism legitimates a flirtation that might be otherwise transgressive; the attraction of objects and the attraction of people elide. Much as social Darwinism would later be distilled from some of Charles Darwin's key concepts—such as natural selection and competition—to describe purely social, not biological interactions, a social distillation of Newtonianism emerges here. What could be called social Newtonianism resembles a loose interplay of attraction—evidenced in the relationship between the narrator and marquise—and the act of looking, two social phenomena that connect to scientific knowledge about physics and optics.

Algarotti's text includes scientific information for those who wished to explore beyond the flirtations or bypass the text's flirtations entirely. The scientific passages connect directly to Newton's research and are signposted dramatically by the text's narrator: "I will not defer this explication any longer, answered I, and I shall be extremely happy if my shewing you, in what manner you see me, may induce you to look upon me in a different way from what you have hitherto done."[50] This signaling would alert those who were only interested in the flirtation between the narrator and the marquise to skip ahead, just as the signal phrase would attract the attention of those who loathed romance and wanted to use the book to teach themselves Newtonian science. The specific language of the translation even implies that revealing scientific information somehow requires the marquise to look at the narrator differently, as if this new information will create new ties between them. With the text's other links between sight and attraction, new scientific insights might indeed draw the narrator and marquise closer to one another.

The narrator differentiates reflection and refraction precisely and correctly. "Light is principally subjected to the two accidents of reflexion and refraction. Reflexion, according to the Cartesians, happens when, by a collision of the globules of light, with the solid parts of bodies, these globules are repelled back again just as a ball rebounds."[51] Algarotti had a comprehensive understanding of Newtonian natural philosophy and performed experiments that followed Newton's methods exactly. Both men and women needed a more comprehensive understanding of Newton's contributions to science, and *Newtonianism for Ladies* helped to provide it. The text's hybridity as part science and part romance meant it was presented to readers in divergent ways. This hybridity is lost if one considers Carter's English translation to be the sole edition.

1739 and 1742: From Translation to Hybrid Piracy

Publishers of the eighteenth-century editions of Algarotti's text engaged readers whose needs varied, providing multiple versions of Newtonian popularizations and narrations of Newton as a figure while circulating under the same author's name. From its 1739 first English-language edition, through midcentury editions in London and in Glasgow, to a 1772 edition, one Italian book inspired by Newtonian natural philosophy became three entirely different English books and one Scottish book. I will describe the books and analyze some essential discrepancies among them, focusing on differences in their intended audiences and differences in how they interpret the concept of translation, the balance between science and entertainment, and the intellectual property of the translator. These editions show that even a single popularization of Newtonian natural philosophy can offer extensive variety.

The 1739 edition, printed by Edward Cave, was the first edition published in England and had a dozen booksellers listed on its title page, so its publication was a major event. One significant aspect of its publication is that a woman, Elizabeth Carter, translated it. Carter, a polymath who corresponded with Samuel Johnson in Greek, remains an interesting figure independent of this study.[52] Mirella Agorni writes that Behn's translation of Fontenelle's *Entretiens sur la pluralité des mondes* "offered her a better opportunity to voice her experience as a woman and a writer than a similar translation would do for Elizabeth Carter fifty years later, in spite of the fact that Carter's work was specifically addressed to a female readership."[53] Agorni argues that Carter's lack of a preface reflects her lack of opportunity, compared to Behn's extensive preface to her own translation of Fontenelle.[54] Cave, the publisher of Carter's translation, also published the *Gentleman's Magazine*, a popular periodical that helped to disseminate intellectual and philosophical ideas publicly. Like the *Gentleman's Magazine*—which was not for male readers alone—*Newtonianism for Ladies* aimed to disseminate complex ideas to inexpert readers, *both* male and female.[55]

The English translations of Algarotti were published in portable duodecimo editions, and were similar in size to novels. These small volumes that could fit a woman's hand were also small enough for men to stash in a corner. Perhaps because the original title referred to "ladies," these books' bindings yielded only vague clues as to their contents. A copy of the 1739 edition bound in 1794 has the binding "Newton on light and colors," and a copy of the 1765 has "Newton philos." Neither binding refers to women as the audience of the texts; in fact, from these bindings a reader might assume Newton had written both works.[56] Hutton describes the use of gender-neutral titling in two 1752 Italian editions of *Il Newtonianismo per le dame:* the work was retitled as *Dialoghi sopra l'ottica neutoniana* (*Dialogues on Newtonian optics*) and *Dialoghi sopra la luce, i colori, e l'attrazione* (*Dialogues on light, colors, and attraction*). Hutton connects these re-titlings to Algarotti himself; Algarotti of course had no control over the English readers who chose to bind their books in a gender-neutral fashion. These English bindings resemble the glossed library titles Moretti describes, in which titles were abbreviated to give readers clues to their contents, because of "special 'signals' [that] place books in the right market niche."[57] The clues on these bindings were not fully indicative of their contents and in fact obscured the fact that the books were mildly sexy popularizations of Newtonian natural philosophy, written by an Italian author and translated by a woman.[58]

The editions of *Newtonianism for Ladies* show that popular Newtonianism included the circulation of pirated works as well as original texts. The second English-language edition, published by George Hawkins in 1742, adapts Carter's translation loosely and interweaves it with a translation of Louis-Adrien Duperron de Castera's French translation from 1738. Some of the body text was translated from the French in 1742, but the footnotes were lifted directly from the 1739 English translation and drew identical comparisons even though many other changes were made to the body text. Even the same items in Carter's translation were footnoted in the 1742 edition. This would not happen in a truly new translation of a text, where an editor would review the entire work for clarity and footnote accordingly.[59] Indeed, both editions include the same unusual imbalance of footnotes: the 1739 has eighteen items footnoted in dialogue 1, and five total footnotes in the other five dialogues. The 1742 edition has eleven footnotes in dialogue 1, all of which are present in the 1739 edition, and five total footnotes—all the same entries—across the other five dialogues.

The use of footnotes in these two editions yields fruitful discussion aside from the piracy issues I address. The insistence on extensive footnoting early on, with little footnoting from the second dialogue forward, evokes recent scholarly analysis on the nature and uses of footnotes in intellectual history. Anthony Grafton's *The Footnote: A Curious History* historicizes the use of footnotes and finds the origin of the modern scholarly footnote in eighteenth-century historical citation practices, so Algarotti's works were published at a time when these scholarly practices were emerging but were not fully codified. Therefore, the uneven citation of material across the dialogues is unsurprising. More recently, Andrew Piper's discussion of footnoting practices mentions the potential for footnotes to evoke a system of other texts—as "print that points to print"—rather than to clarify or define terms.[60] *Newtonianism for Ladies* contains both kinds of footnotes, including citations for historical figures such as Galileo and citations for books such as "Mr. Swinden['s] . . . Treatise *Of the Nature and Place of Hell.*"

A rare exception to the cuttings and pastings in the 1742 edition's footnotes is a footnoted reference to the eminent early modern historian Francesco Guicciardini. The 1739 Carter translation refers to Guicciardini as "an Italian historian very prolix and tedious." In 1742 he is described differently—with similar disdain—as "a very tedious Italian historian." Prolixity appeared to be a quality that Duperron de Castera and the 1742 English translator of his French translation much admired, as will be seen. Most likely the 1742's footnote to Guicciardini stemmed from Duperron de Castera's French, where "Guichardin['s]" work is an "ouvrage Italien extrémement ennuyeux," emphasizing the extent of his dullness.[61]

This kind of editorial commentary was among the reasons Algarotti himself disliked Duperron de Castera's translation of his work; Algarotti wrote to Voltaire in 1738 that he had translated two of *Il Newtonianismo per le dame*'s dialogues himself from Italian to French in an effort to correct the translation. Algarotti wrote, "La meilleure façon de lui faire voir toutes ses fautes, c'est, je crois, de faire une autre Traduction," claiming, rather generously, that translating the dialogues for Duperron de Castera would show him the faults of his translation. This is a lesson much in keeping with *Il Newtonianismo per le dame*'s didactic aims.[62] It is easy to see why Algarotti found Duperron de Castera's work unacceptable: its footnotes actually undermined Algarotti's writing at times. In a footnote, Duperron de Castera interjects, "* Que veut dire la marquise? Est-il bien vrai qu'en général les Poëtes soient plus doux & plus moderés dans leurs Ouvrages que dans leur interieur? On a beaucoup de preuves du contraire."[63] With these lines, Duperron de Castera inserts *himself* into Algarotti's text—"What is the marquise saying?"—then offers his own baited question and answer: "Is it really true that in general Poets are softer and more moderated in their Works than in their interior? We have much proof of the contrary." The translator's profoundly unnecessary attempt to re-mediate Algarotti's framing of Newtonian natural philosophy to make a point about the sensitivity of poets must have been irksome to Algarotti, whose French was indeed good enough for him to be able to translate well. Voltaire noted Algarotti's skill with languages after an early meeting between the two men in 1735.[64] *Il Newtonianismo per le dame* has additional links to French culture beyond the translation of Duperron de Castera and Algarotti's relationship to Voltaire; other scholars have suggested that the marquise was modeled on Emilie du Châtelet, who was close to Voltaire and who knew Algarotti.

In addition to revealing the popularity of Newtonianism among English readers, the 1742 edition also presents a unique case in the history of eighteenth-century intellectual property. It illuminates the way intellectual property was viewed in new translations and editions. The footnotes to a work that identify or explain some of the cultural references present in the original language, even though many of them require more research and labor than translating text alone, do not seem to be regarded as another person's intellectual property in the same way as the body text. The addition of translations from the French in important places, especially the beginning of the book, was most likely intended to help Hawkins and company avoid accusations of copyright infringement. English copyright laws under the Statute of Anne respected the rights of translators, and Carter and Cave could have pursued legal action against Hawkins and the translator, yet did not do so.[65] Even if Carter knew about the piracy, some might have felt it improper for a woman to claim as her own intellectual property her translation of a man's text—especially a man who was still living and writing. Algarotti himself was buying and collecting paintings while this piracy was circulating, and these other pursuits may have distracted him from tracking down possible misappropriations of his text.[66] Considering his low opinion of the French translation, it is unlikely that he would have heralded its further translation into English and its merger with an English one.

The identical footnotes taken from Carter's 1739 edition are distinct from the many liberties that the 1738 publication of Duperron de Castera's translation—and, following it, the English 1742 edition—takes with the text of the original Italian. These contrasts are evident throughout the text. To illustrate the extent to which the 1742 edition embellishes the language of the 1739 translation, compare a line from the beginning of the text in

Algarotti's 1737 edition to Carter's first English translation. Algarotti's text begins, "Quella stessa ragione che ad un concerto di musica, a una allegra e delicata cena, a una galleria, o ad un teatro tutto dì mi conduce, quella stessa a scriver."[67] Carter's translation begins, "The very same reason that led me every day to a concert of music, a gay and elegant entertainment, a ball, or the theatre, induced me to write."[68] The two share syntax, meaning, vocabulary, and usage. Readers would have expected the 1742 edition to derive from the same exact base passage in Italian. Even the more recent 1739 updated Italian edition began in the same way. The 1742 edition begins differently: "Pleasure is the general motive of our actions: we search it in the cabinets of the curious; at consorts of musick, at plays, and publick entertainments; it leads us to assemblies of gaiety and politeness. 'Tis my inducement too, to give an account."[69] The extensive writerly elaboration does not come from the Italian original. Duperron de Castera's 1738 French translation, however, begins, "C'est le plaisir qui nous conduit à une galerie curieuse, au concert, au spectacle, à un festin ordonné par la joye & par la délicatesse. C'est aussi le plaisir qui m'engage à raconter une histoire."[70] The introduction of the word "pleasure" ("plaisir"), the addition of "cabinets of curiosity," and the addition of "assemblies" after the reference to the theater all point to the 1742 English as a descendant of the 1738 French translation. Agorni describes Carter's translation as faithful but conservative, removing references to sperm and downplaying references to the marquise's body; Carter would not have approved of the unnecessary elaborations that the 1742 makes to the text, even though Agorni, like most scholars, describes the 1742 edition as a reprint.[71]

In the 1739 English edition, which remains faithful to the Italian, the text is introduced as "an account of the manner in which I passed my time last summer in the country with the Marchioness of E___, and has thus from an idle and useless member of society, rendered me an author."[72] In 1742, half of this sentence becomes twice as long: ". . . Marchioness of ______. As I happened, during the whole time of our Stay, to be very much at leisure, free from the noise and perplexity of business, the calmness of so agreeable a retreat presented me with a fair opportunity of turning author."[73] Not only is the narrator transformed into a respectable businessman, instead of "an idle and useless member of society," but these lines also make authorship seem like a labored career path, not just having pleasant conversations and writing them down.

This sympathy with creative work aligns with Duperron de Castera's defense of poets in the footnote on page 16, discussed earlier. The 1738 French translation champions the "belle solitude" of the countryside; Duperron de Castera's use of the word "désoeuvré," which can mean "jobless" as well as "idle," explains the description of a writerly occupation.[74] Following the 1738 translation's form, the 1742 English edition breaks into another paragraph to expostulate on the connections between authors and publicity, whereas Carter's translation and the 1737 Italian edition do not. These embellishments reflect the influence of popular texts like novels on popular prose and changing social perceptions of career paths, suggesting writing as a legitimate career path rather than the hobby of the idle rich.[75] Duperron de Castera also wrote fiction, which also accounts for the text's embellishments to Algarotti's original Italian.[76] Eliza Haywood translated one of Duperron de Castera's novels into *The Lady's Philosopher's Stone* (1725), a connection that further exemplifies the slippage between science and fiction during the eighteenth century.

Other uses of the 1738 French translation proliferate in the 1742 edition; some give the character of the marquise—the marchioness in English editions—a more fully developed presence in the text. Agorni describes Algarotti's work as a "masculine discourse relegating woman to her traditional position as object," and indeed the early editions, including Carter's translation, offer the marquise little opportunity to do more than query the narrator.[77] Even though I would qualify the idea that this text is as "traditional" in its depiction of women as the thoroughly objectified feminine entities that pass as women in Halley's and Thomson's poetry, Agorni is correct to identify that the 1739 translation is very light on what might be called "feminine discourse." The marquise actually speaks more in the 1742 edition. The entertaining aspects of Algarotti's work seem to have inspired translators like Duperron de Castera, who took advantage of the kind of storytelling present in novels and added dimensionality to her character. In this translation, the interactions between the narrator and the marchioness seem less like he is teaching her and more like they are interacting with one another. In Algarotti's 1737 and Carter's 1739 texts, the marchioness's character had existed to enable the narrator to tell his story, one possible side effect of its being a text that used conversation to instruct. She is permitted the occasional interjection, such as "'I see,' said the marchioness, interrupting me," while the narrator is teaching her.[78] The 1742 edition is still a text that teaches Newtonian natural philosophy, but the marchioness's response to the same prompt differs, changing to "'Hold, hold, dear sir!'

said the marchioness, disappointing me."[79] Her character is assertive and lively during these passages, but the narrator's disappointment at her assertiveness means she does not fully escape criticism. The 1739 edition silently communicates the marchioness's perplexity over the words "sevenfold light and Newtonian picture."[80] In 1742, following Duperron de Castera's French translation, she speaks in response to this perplexity, crying, "What cramp words are here!" when confronted with the same language, using slang inappropriate to her social rank.[81] Duperron de Castera had indeed footnoted the term for "sevenfold light" as an especially challenging note in his translation. Although this passage indicates that the embellishments to the text lead to a more prominent role for the marchioness—who communicates her opinions more actively—she is more overtly sexualized in this edition, as my subsequent analysis reveals.[82] This ambivalence about women's assertiveness is a surprising turn in a work that aimed to teach science to women, and it facilitated the work's reappropriation in more gender-neutral contexts.

The two editions diverge further in an important passage for understanding the Newton-Descartes conflict identified earlier in this chapter, when Algarotti shows how Newton differed from Descartes. English readers needed this discursive contextualization to demonstrate competence in natural philosophy, so it is essential in this passage to translate Newton to the English. Algarotti refers to Descartes throughout *Il Newtonianismo per le dame*, refraining successfully from slighting the French philosopher even though this book focuses on the "deity" Newton. Fontenelle's *Entretiens sur la pluralité des mondes* had popularized Descartes in late seventeenth-century Europe, and Algarotti's work responded thematically as well as scientifically to Fontenelle's. Algarotti knew that his audience might have questions about how Newton expanded on Descartes's contributions to science. In the first dialogue, Algarotti connects Newton's discoveries to those of Descartes in a passage on the importance of foundational systems to scientific study: "How then could Descartes for example, who was the chief author of this enterprising sect of philosophers, make a rational system concerning light and colors, when he was entirely ignorant of so many of their qualities which Sir Isaac Newton afterwards discovered from observations? How could he form the statue, when he had not the marble?"[83] Algarotti suggests that Descartes might have made conclusions similar to Newton's if Newton's data had been available to him, simultaneously offering respect to Descartes as a fellow "artist" and crediting Newton with scientific innovation.

In the 1739 translation, the marchioness also explains Newton's appeal to women, and to audiences of popular literature: "I, who am a woman, replied the Marchioness, confess that I love those who have the courage to venture upon grand and difficult enterprises. Is not this the reason why we interest ourselves so much in the adventures of heroes? The courage of those philosophical heroes has something sublime and superior in it: if they do not attain to all they promise, must there not be given some indulgence to the imperfections of human nature?"[84] The first quality of this "philosophical hero" that impresses the marchioness is Newton's bravery. Like the masculine Newton who metaphorically conquered the earth in poems by Thomson and Halley, this Newton is also a hero because he attempts challenges on a grand scale. These depictions could be foils for the divine simplicity of Alexander Pope's famous couplet memorializing Newton, in which "God said, 'Let Newton Be!' and all was light."[85] The marchioness refers to Newton's courage as "sublime and superior," defining him as an exceptional hero whose bravery is as important as the knowledge he illuminates.

The references to Descartes and the discussion of Newton's appeal to women differ in the 1742 edition. Because the 1742 edition was in part modeled on a French translation, it affords more space to Descartes, granting Descartes his own section, and adapting the text of the 1739 translation: "Descartes, for instance, who was the chief author of this enterprising sect, how was it possible for him to form any rational system of light and colours, ignorant as he was, of their principal properties? and which would have been still a secret, had not Newton published his discoveries. Was not this attempting to carve a statue without the marble?"[86] The 1742 text responds to Newton as an author whose substantial contribution to science is that he "published his discoveries," even though Descartes is referred to as the "chief author" of seventeenth-century natural philosophy. The translator implies that Descartes possessed interpretive skills that were as fine as Newton's: the ability to carve the statue is not questioned. Newton, as the person who unglamorously provided the marble, was able to move to the forefront of science.

Marble-related metaphors in *The Mind Is a Metaphor* include marble hearts, marble eyes, limbs that are stiff like marble, and—with a metaphor similar to what Algarotti is using—the connection between knowing what to do with a block of marble and knowing what to do with one's expertise.[87] The closest the database can give us to a connection between marble and print is a Leibnizian comparison, derived from Locke: "I have also used the analogy of a veined block of marble, as opposed to an entirely homogenous block of marble, or to a blank tablet—what the philosophers call a *tabula rasa*."[88] Apart from the potential for a block of marble to be inscribed—as in, when we are looking at it from its own perspective rather than that of the expert—few examples compare the materiality of marble and print publication, even though that is the reference Algarotti makes.[89]

Duperron de Castera attempts to balance the credit due to Newton and Descartes, but his translation actually takes the passage in a different direction from Algarotti's claims. Considered thus, the 1742 edition offers the English an altered perspective on the relationship between Newton and Descartes, largely because it is modeled on a French translation for a French audience. If they had read the 1742 edition, the English champions of Newton who were described by Voltaire may have given Descartes more advantage in their arguments.

The 1738 French and 1742 English texts also adapt the marchioness's response to Newton's philosophy: "I must own, for my part, said the Marchioness, I have thus much of the woman in me, that I admire and respect men who undertake great and difficult enterprises. Is it not merely on this account that we so deeply interest ourselves in the adventures of heroes? There is a noble air of superiority in the daring Genius of these philosophic worthies, which gives me vast pleasure: and if they do not always keep strictly to their word, we should make allowances for human weakness."[90] In this passage, the marchioness places additional emphasis on her gender, calling into question the seeming independence she had displayed in the earlier interactions between her and the narrator. She quantifies her gender, claiming that she has "much of the woman" in her, and emphasizes the fact that men assay the "greater and difficult enterprises" in particular. Her admiration of the "noble air" of philosophers and the "vast pleasure" they give her makes natural philosophy itself seem erotic. She does not, however, refer to Newton by name in this passage, as the 1739 edition had done. By addressing men as a general category, the marchioness makes Newton seem like one of a type of men. Following this logic, Newton and Descartes may be considered equals, much as their skills are ranked comparably in the prior passage. The increasingly erotic tone of this passage alters the tone of its conclusion. The disclosure that "we should make allowances for human weakness" if men "do not always keep strictly to their word" carries an erotic charge on top of its literal reference to Descartes.[91] Here, as in Halley's and Thomson's poetry, intellectual achievements are both sexualized and masculine. Kathleen Lubey writes of explicitly erotic texts that such works "explicitly theorize the imagination as a faculty whose pleasures can be both enjoyed and trained to serve instructive ends."[92] The presence of erotic elements in this instructive, co-ed text does not therefore surprise.

Readers thus learned different information about Descartes and Newton from the 1742 edition than the 1739. Even from the beginning of Duperron de Castera's French translation, the relationship to Descartes was foregrounded. The 1738 French edition's title page placed the subtitle *Entretiens sur la lumière, sur les couleurs, et sur l'attraction* beneath Algarotti's title—a reference to Fontenelle's *Entretiens sur la pluralité des mondes* and a nod to Descartes, whose natural philosophical principles Fontenelle had shared in conversational form. If anything, the subtitle of Algarotti's work appears more diffuse and less entertaining than Fontenelle's *Entretiens*.

Michèle Cohen describes eighteenth-century French textbooks used by the English in the eighteenth century, which move from using conversation as an educational tool to using grammar as an educational tool. The use of conversation as a teaching tool correspondingly diminished in the shadow of grammatical instruction. This trend may in part explain Duperron de Castera's adaptation of conversations in Algarotti's text into a more novelistic format: if the use of conversation for didactic purposes was dying out, conversation could still be used as a form of entertainment as it was in the dialogue of novels. As the century progressed, sexuality was also represented in popular works addressed to both men and women, often in the form of letters—a form that related to written conversations as well as the novel. As Ruth Perry describes, readers of both sexes enjoyed "endless discussions of the proper levels of prudery and sociability with anecdotes about flirtation and seduction."[93] These conversations were thus part of a general trend toward instruction that did not sideline pleasure as a new category of teaching text emerged during the century. Algarotti would later produce other entries in this field, including his essays on opera and painting discussed later.

Not all of the 1738 French and 1742 English embellishments to the text increase the presence of the marchioness's character or the text's novel-like qualities. Some of the amendments improve the reader's understanding of its content. In all editions, the narrator reads a selection from "Ode to St. Cecelia's Day" to the marchioness—not the Dryden version, but that of Alexander Pope, still living in 1742. In the 1739 edition, Carter cites lines 10 and 11 from the poem, "While in more lengthen'd notes and slow, / The deep, majestic, solemn organs blow" and, confusingly, includes Algarotti's translation of the lines into Italian: "Mentre con tarde ed allungate note / Il profondo, solenne, e maestoso / Organo soffia."[94] On the next page, the narrator and the marchioness discuss a section of an Italian poem—written by Algarotti himself—that puzzles the marchioness, and it is rendered appropriately, in English with an Italian footnote. The 1742 edition adds additional lines from Pope's poem, providing twelve lines of the poem instead of two, in order to provide more context, ending as "the strains decay, / And melt away, / In a dying, dying fall."[95] The 1742 edition improved the text by adding additional lines from Pope's poem and by removing the redundant lines of Italian translation.

Although the 1742 English edition borrows in many places from the 1738 French edition, the deeper one reads, the more the 1742 English edition resembles an exact copy of Carter's 1739 translation. This is another reason I default to Carter's translation in my close readings of the eighteenth-century English text. For example, the quotations from dialogue 2 that were discussed earlier in this essay are identical in all English-language editions of Algarotti. Even though the embellishments to the text only affected the portions of the work that were immediately apparent to a casual reader, such as the beginnings and ends of sections, the 1742 edition looked sufficiently distinct from Cave and Carter's version that critics have not noted this piracy and in fact may not have even read more than one of these editions. The 1742 edition of *Newtonianism for Ladies* exemplifies the dangers of failing to discriminate among editions in scholarship as well as the ways eighteenth-century publishers circumvented copyright laws. Far from being excluded in favor of Carter's translation, the 1742 edition should be valued on its own as a hybrid work that offered a more fully developed female protagonist, raised the profile of Descartes in eighteenth-century England, and narrowed the gap between popular science and popular fiction.

1765 and 1772: Addressing Diverse Readers

Later editions of *Newtonianism for Ladies* published in 1765 and 1772 reveal publishers who address different expectations among a gendered readership of both women and men. For example, the next edition, printed in Glasgow in 1765 by Robert Urie, makes its own modifications to the 1739 Carter translation that change its market positioning and its implied use from a gendered to a gender-neutral—perhaps even overtly masculine—

readership. Replacing the 1739 English title, which read *Sir Isaac Newton's Philosophy Explain'd For the Use of the Ladies,* the 1765 edition reads *The Philosophy Of Sir Isaac Newton Explained.* Newton is explained, without explicitly addressing the ladies, in the same six dialogues as before.[96] The listing of Algarotti as author also differs in the 1765 Glasgow version: instead of "from the Italian of Signor Algarotti," as the 1739 edition familiarly describes the author, the 1765 considers him more formally, as "Count Algarotti, FRS, FSA."[97] This formality may have been used because Algarotti had been awarded membership in these learned British societies; his status as a Fellow of the Royal Society and Society of Antiquaries emphasized his status as an expert, rather than a learned amateur.[98] Such slight alterations in the Glasgow edition thus work to elevate the text's and the author's profiles for a broader audience: male or female, with a serious interest in science and culture. Unlike the 1742 edition of *Newtonianism for Ladies,* the 1765 edition remains faithful to Carter's 1739 translation. Even though this edition nearly exactly replicates Carter's translation, the work's paratexts make it appear as gender-neutral as possible without changing the document's content.[99] Line by line, apart from an occasional variant spelling that may have come from typesetting, the body text of the 1765 edition is identical to Carter's translation.

Additional elements besides the 1765 edition's gender-neutral title mark the work's positioning for a less feminized audience, including the disappearance of its epigraph. On the title page of Algarotti's original text and the 1739 edition is an epigraph from Virgil's *Eclogues,* "Quae legat ipsa Lycoris," which translates to "Lycoris herself may read." All other English editions of the book use this epigraph, which provides a direct address to a literate female readership and represents an appeal to historical texts for evidence of literate women. The epigraph would have been an important component of marketing the book to women. Because Virgil's Lycoris was also a courtesan, this epigraph gestures to the book's more titillating sexual content. The woman who pursues potentially transgressive knowledge may be open to transgressions of another kind. Agorni describes Algarotti's book as "far from recommended to a female readership" because of its objectification of women.[100] The address to Lycoris implies that it did gesture to a female readership, a readership with some sophistication regarding sexuality. If readers of *Newtonianism for Ladies* were granted erotic maturity alongside scientific sophistication, removing the epigraph might cloak this transgressive gesture. The epigraph disappears from the 1765 edition, showing that Urie and company manipulated the form of the text to present it to a different readership, even though the language of the body text itself remained the same. The epigraph is present in the 1742 and 1772 editions, which were more explicitly marketed to women in other ways. The text's appeal to a different readership is thus marked by the absence of a female audience rather than the explicit presence of a masculine audience. In this instance, what we may describe as gender neutrality today may simply have meant masculinity in 1765.

The 1765 edition also uses far fewer typographical distinctions within its body than the 1739 work from which it borrows. In some sections the lack of typographical variation appears to be a strategic effort to make the text seem more unisex. An example of typographical distinction from the 1739 edition is a line from its introduction that describes the reasons behind popularizing science; it is a topic "where ladies would rather *perceive* than *understand*." This emphasis, likely intended to draw out thinking about the nature of the words "perceive" and "understand," is missing in the 1765 version, possibly because the reference to a female audience was less important. Another possible explanation is that a typesetter chose not to use italics. Yet the differences between the appearance of this text and that of the 1739 edition are substantial, and these differences include typographical variation, whatever the reason behind it. They would not have been sold as the same book in the same way, even though the information contained in the texts is identical.

The books that were advertised in this edition also gesture to the implied audience for the text. Among the books advertised in Urie's 1765 edition is Algarotti's *Essay on Painting*, discussed later in this chapter, as well as other works in translation by D'Alembert, Huygens, and Voltaire. According to Richard Sher, Urie worked mostly in reprints and books in translation.[101] Most of his titles are for self-improvement and education, another factor that accounts for the seller's downplaying women as an audience. Urie was likely attempting to achieve an eighteenth-century form of brand consistency because he catered primarily to the self-improvement of men rather than women.

The forty-five years separating the death of Newton and the publication of the 1772 edition, printed for Francis Newbery, had seen much growth in amateur scientific readership. Voltaire had lamented in the 1730s that the English attempted to win philosophical debates about scientific works they had not read; by 1772 this practice had become a part of the circulation of ideas, with an entire industry of works designed to translate science to nonprofessional readers.[102] The 1772 edition of Algarotti's text has a new title that is far different from that of the 1739 text. Its title is *The Lady's Philosophy: or Sir Isaac Newton's Theory of Light and Colours, and His Principle of Attraction, Made Familiar to the Ladies in Several Entertainments. A New Edition. Translated from the Original Italian of Signor Algarotti.*[103] The new English title is longer and far more explicitly marketed to women than the titles of earlier editions. The book is described as *The Lady's Philosophy* rather than Newton's. It specifies the principle of attraction as an important component of popular Newtonianism and refers to the dialogues as entertainments. It is also labeled as "new."

Readers may question my comparison of *The Lady's Philosophy*, which bills itself as having come from the original Italian, to Carter's translation. The translation is not, however, original. In fact, it is a direct copy of the 1742 edition: the hybrid of the 1739 translation by Carter and the 1738 French translation. It uses as its source text an already pirated translation. As with the other English-language editions, *The Lady's Philosophy* makes no reference to the lady who initially translated it.[104] I am again indebted to Moretti's discussion of the importance of glossed or abbreviated titles in the reception of a work: to describe this work as *The Lady's Philosophy* is to imagine an entirely new work from a marketing and sales perspective, regardless of the similarity of the actual text to other editions. Describing it as "the" lady's philosophy implies that one, knowable philosophy for all ladies is within this edition. The 1772 edition's title refers to "attraction," "familiarity," and "entertainment" as related to its "lady's philosophy," suggesting the flirtatiousness with which Duperron de Castera and the 1742 translator embellished Algarotti's content. This edition normalizes a pirated text by using it as a source, falsely aligning the words "new edition" and "translated from the Italian" as if to imply that it is both a new edition and translation. Nevertheless, Newbery's readers were unknowingly purchasing copies of a pirated translation that had been published in 1742.

Newbery published popular scientific works and differentiated his titles from others on the market. His father, John Newbery, had published the children's science book series whose full title was *The Newtonian System of Philosophy Adapted to the Capacity of Young Gentlemen and Ladies*, beginning in 1761, and Francis Newbery continued in the family business after his father's death.[105] *Tom Telescope* may help to account for the titling of *The Lady's Philosophy:* the young men in its audience would graduate to reading Newton himself or other primary sources, and the young ladies could graduate to *The Lady's Philosophy*. Benjamin Martin's *The Young Gentleman and Lady's Philosophy*, printed and sold by William Owen as well as Martin himself, offered similar benefits to late eighteenth-century juvenile readers; it included instructions for the proper reading of a globe and advanced to the basic use of optical instruments.[106] Martin's readers, like *Tom Telescope*'s, were well prepared to understand basic scientific experiments as well as the natural phenomena surrounding them. "J. Newbery and J. Coote" are listed as having printed *The Lady's Museum*, by Charlotte Lennox, the author of *The Female Quixote*.[107] So calling this work *The Lady's Philosophy* also aligned it nominally with a different kind of work, a periodical by a female writer that was addressed to women. *The Lady's Museum* has sections on natural history described as "Philosophy for the Ladies." An edition of popular Newtonianism aimed at an adult audience like *Newtonianism for Ladies* was the right kind of text to add to the Newbery catalogue: in addition to the young ladies who grew up to become adult readers, the mothers of the children who would read *Tom Telescope* could purchase a corresponding volume for themselves and enjoy the diversion of a titillating courtship story. The varied popular science titles published in one house—for children and for adult women—reveal how diverse the field of scientific publishing had become by the last quarter of the eighteenth century.

Newbery's works have attracted the interest of critics like James A. Secord and Pete Langman, who both describe the popularity of *Tom Telescope*. Langman in particular analyzes the ways that *Newtonianism for Ladies* translated virtual witnessing into fictional terms in order to replicate experiments for readers, writing that "*Newton for the Ladies* seems to have been edited in order to accentuate its poetic nature, a nature designed to conjure images in the imagination of its readers."[108] In light of my earlier analysis of Newton's—and the *Principia*'s—resistance to virtual witnessing, it would follow logically that a popularization like Algarotti's could be effective by reframing the narrative as a virtual witnessing experience. Especially in England, where Newton's was a household name that represented the heights of English rational achievement, purchasing a text about Newton offered people from diverse social ranks opportunities to forge unique connections to the intellectual productions of their nation. *Newtonianism for Ladies* has not only a "poetic nature," as Langman describes, but a prosaic nature as well, its different editions paralleling the trajectory of early novels that dramatized flirtatious interactions between men and women.

In their work on *Tom Telescope*, Mary Fissell and Roger Cooter describe the extent to which it "adopted many of the forms in which Newtonianisms were being presented in eighteenth-century Britain"; considering these multiple Newtonianisms brings us closer to understanding who Newton was to English readers at the end of the eighteenth century.[109] In the concluding chapter to this book, I will examine evidence of an even broader readership for popular Newtonianism at the New York Society Library. In Britain in the 1770s, there were Newtonianisms for amateur ladies, for gentlemen, and for children as well as for a more scientific and academic audience.[110] Simply mentioning the existence of Algarotti's text and providing a brief summary, as most critics do who focus on the popularization of science in England, does not suffice. We must account for the diverse ways *Newtonianism for Ladies* popularized Newton, because even this single popularization had significant commercial and entertainment factors to be considered that varied from edition to edition. Although the multiple English editions of *Il Newtonianismo per le dame* contain less "hard science" than Newton's own works and other works of Newtonianism, the editions I have described in this chapter are nonetheless essential to understanding the popularization of science in England during the eighteenth century. The study of *Newtonianism for Ladies* also presents significant implications for scholarship of eighteenth-century publishing practices, particularly the orientation of publications toward a gendered print market. Finally, the study of English-language editions of Algarotti's text further destabilizes the already blurry distinctions between science and literature by acknowledging such variations across a single title. Algarotti's popularization of scientific ideas was even more complex than this—it included the presence of Newtonian natural philosophy and popular science more generally across Algarotti's English-language works.

Complicating Popularization: Scientific Content in Algarotti's *Essay on Opera* and *Essay on Painting*

The complex relation of gender to audience in *Newtonianism for Ladies* is more diffuse than the explicitly defined masculine audience for the print communities that encountered Algarotti's *Essay on Opera* and the *Essay on Painting*. The gendered nature of the reading publics whom eighteenth-century publishers addressed in diverse editions of *Newtonianism for Ladies* meant shifts in tone depending on the market's gender. Changes in editorial and print practices enabled these shifts, and reading publics and reading markets had not yet consolidated. The following section engages with interdisciplinary texts with seemingly diverse markets to situate popular Newtonianism within the discourses of art and music as well as popular science. Carolyn Guile shows how Algarotti "used the terms of Newtonian science to forge a . . . sense of temporal unity between the art of the pre-Newtonian past and that of his own day.[111] Algarotti made essential links between art and science that reveal the extent to which he helped spread scientific ideas across Britain and Europe.

In addition to being a Fellow of the English Royal Society and acquaintance of Voltaire, Algarotti was also a fixture at the court of Frederick the Great and a supporter and critic of the arts. Stanford's online project *Mapping the Republic of Letters* reveals the extensive reach Algarotti and his work had in Europe.[112] *Il Newtonianismo per le dame* helped spread Algarotti's reputation across continental Europe in print; his published writing gave readers some information about Newton's work that could be assimilated without an extensive knowledge of mathematics. *Newtonianism for Ladies* as a standalone title, or series of editions, does not tell the whole story that connects Algarotti's work to English readers of the eighteenth century; this cultural transmission includes Algarotti's other works, including his popular essays on painting and opera, and we gain a great deal from considering these works together as a small but significant part of reading popular science in England. As Guile writes, Algarotti helped "'Venetianness' in painting" to "acquire a scientific dimension," with his connection of Italian painting techniques to Newtonian optical theories.[113]

These works, *Saggio sopra l'opera in musica* (*Essay on Opera*, 1755) and *Saggio sopra la pittura* (*Essay on Painting*, 1762) actually contain some scientific content that English readers would have encountered. These texts thus support the interconnectedness of science and the arts, much as *Newtonianism for Ladies* had done starting in 1737. Each of the essays connects good art to good science, and poor art to poor science. The essays on opera and painting also function differently than *Newtonianism for Ladies* with regard to gender. Algarotti had addressed women explicitly in that title, whose women readers learned about science in a format that was combined with the rhetoric of artistic appreciation. In the essays on painting and opera, an implied male audience learned artistic content with scientific content included. This narrative hybridity, this interplay of scientific terminology and artistic forms, oriented English readers to a Newtonian perspective in arts and sciences. These three texts in England, then, form a network of hybrid teaching texts that familiarized English readers with scientific concepts. The publishing history of the essays on painting and opera supports the idea that their audiences overlapped. They did not publish *Newtonianism for Ladies,* but the London publishers Davis and Reymers published English editions of both the painting and opera essays in 1764 and 1767.

Some have speculated about who the anonymous translator of the essays on painting and opera was and whether it might be Elizabeth Carter, the translator of the 1739 *Il Newtonianismo per le dame.* It may be possible to tell if Carter translated it using text analytics, since she previously translated a work by Algarotti; however, a positive result might only reveal that the works are all literal English translations of Algarotti's own writing style, because her translation was very faithful to Algarotti's original Italian. A scholar might, then, only prove that the work was Algarotti's rather than Carter's. Carter was a well-known translator during the latter part of the century, especially for her translation of Epictetus for Samuel Richardson in 1758. That translation had been a high-profile work, with her name on the title page, and a long list of subscribers that included the Prince and Princess Dowager of Wales. It was sold by Andrew Millar, John Rivington, and Robert and J. Dodsley. These sellers all had relationships to different intellectual and imaginative works. Millar had helped finance Johnson's *Dictionary* and paid well for works such as Thomson's *The Seasons,* which also included Thomson's poem on Newton discussed in chapter 3. Rivington's family had sold *Pamela* decades before and had worked closely with Richardson. Dodsley had works by Thomas Gray and Laurence Sterne and also published his own poetry.

Davis and Reymers, while also respectable, printed works like travel writing, miscellanies, and sermons rather than original works, so translating for them may not have been something Carter would have advertised, especially during a time when she was associated with a number of high-profile subscribers. The Davis and Reymers catalogue was eclectic, but it also included John Kennedy's *A Complete System of Astronomical Cosmology* and, as such, had a hand in some forms of scientific popularization (1762).[114] Perhaps more in tune with practical science, Davis and Reymers also published the third edition of Edmund Chapman's *Treatise on the Improvement of Midwifery* (1759), printed by "The Printers to the Royal Society." These publishers may not have been substantially different from other publishers, except for the fact that they were not publishing as many new works, or shaping markets by commissioning new works. Carter's high-profile translation of Epictetus may even have outranked Algarotti's own work by the time the essays on painting and opera were published.

Regardless of the gender of the translator, the essays on painting and opera explicitly courted an audience of men the way that *Il Newtonianismo per le dame* had courted women. One gendered metaphor in the *Essay on Painting* exemplifies the way the essay addresses a larger readership while asserting that only men could pursue painting as a career. In a section on discovering future painters, Algarotti describes his theories of education, which apply to men exclusively. He wishes that princes could

> place men of penetration in the public schools; in order to examine and trace out the various inclinations of the youth who frequent them. By laying before them, from time to time, instruments of mathematics, of war, of music, and in short, of all the other arts and sciences; and by repeated trials of them in the use of these instruments, they should be forced and constrained to manifest their several geniuses in the same manner that the artful Ulysses, by introducing glittering arms as well as precious jewels to the daughters of Lycomedes, found means of discovering Achilles, who lurked amongst them in the habit of a female.[115]

Algarotti codes genius as male, Ulysses using his wit to deceive women with gifts and uncover the hidden Achilles. Achilles's *true* identity—here compared to learning that one is truly a painter—is masculinity that has been cloaked in the habit of a woman. The legend's alignment of feminine appearance with deception, with artifice rather than art, amplifies the distance between the women whom Algarotti engaged in *Il Newtonianismo per le dame* and the men who were the de facto potential artists of the *Essay on Painting*. The women in the legend can be bribed easily, are motivated by material rather than transcendent goals, and are merely instrumental to one man's pursuit of another man. This was not exemplary if a woman wanted to be an artist, or if she wanted to exhibit any form of agency, much like in *Newtonianism for Ladies*, where the implied readership is students of science rather than creators. The passage also gestures to Algarotti's scientific background. Even though it comes from an essay on painting, Algarotti describes scientific and mathematical instruments to which young boys may be exposed, and levels out the differences between these instruments: the term "instrument" connects science, math, war, art, and music, each rendered as a valuable and manly pursuit.

The classical text to which Algarotti refers, Achilles on Skyros, from the *Achilleid*, was a popular subject of operas, so Algarotti would have known it in different iterations. His essay on opera also mentions the work, as part of a compliment to the librettist Metastasio for choosing such an inspired subject.[116] *Newtonianism for Ladies* refers to Achilles in dialogue 3, where the marchioness and the narrator look at a painting of Achilles in her gallery, and the narrator argues that Descartes's theories of light are unrealistic: "Place yourself at this Pilaster and look upon the red of the vesture of Achilles. I will stand at the window, and look upon the blue of the Sea, so that while you are looking at the red and I the blue, our Eyes may each be directed to the same Point of Air. It is certain that two rays will pass through this point, one from the Robe of Achilles, and the other from the Sea."[117] If the eyelines were composed of globules of light, the trajectories of the narrator's and marquise's foci would unite. Achilles and the sea are used to explore the attraction between the narrator and the marquise. He asks her, suggestively, to imagine "Two Series or Strings of Globules immediately touching each other, the one continued from the Robe of Achilles to your Eye, and the other from the Sea to mine. . . . There will be one Globule common to both Strings. Do you understand all this?"[118] The marquise replies, ambiguously aroused, "Yes, too well. . . . And I begin already to tremble."[119] In the painting under discussion, Minerva attempts to stay Achilles from killing Agamemnon. The narrator imagines a third spectator eyeing "the golden Colour of Achilles's Hair, that Minerva lays hold on in order to calm his destructive and impetuous Rage."[120] Minerva, most likely a reference to the wisdom-seeking marquise herself, is responsible for soothing and calming men's transgressive physical impulses.[121] The narrator himself postures as if the dialogue has enraged him, claiming, "I beseech you, for the Love of Philosophy, never mention these Globules to me again. I resolve to think no more of them since they have so shamefully yielded to the very first Difficulty. They seem to resemble those unexperienced Lovers, who at the first Repulse think of a Retreat."[122] The tension between the gaze of the marquise and the narrator deflates into a failed seduction attempt; however, the object of the seduction is the narrator's credulity, rather than the marquise's affections. The globules are unworthy suitors to the narrator's intellect, Descartes a weaker seducer than Newton. The presence of sexuality and flirtation in the discussion of Achilles in *Newtonianism for Ladies* stands out against the more transparently allegorical references in the *Essay on Painting* and the *Essay on Opera*. As I have discussed previously with regard to what we would call "gender-neutrality," this neutrality in effect meant "non-feminine," as both of these works appealed to a male audience.

Publishers tried to homogenize all three of these texts by Algarotti as masculine. In Glasgow, all three of these books—the essays on opera and painting as well as *Newtonianism for Ladies*—were printed for Robert Urie. The essay on painting was first, in 1764, the year Algarotti died, followed by *Newtonianism for Ladies* in 1765, the essay on opera in 1768, and Algarotti's letters in 1770. An advertisement in the back of *Newtonianism for Ladies* mentioned the book on painting, so Urie clearly believed that the works would have shared an audience. Earlier I noted that this edition downplayed the gendered aspects of Algarotti's original text, even though the text itself was mostly a reprint of Carter's translation. Comparing this edition with the masculine audience for the other two works, which do not address women except in passing, Urie's 1765 edition had its gendered qualities toned down because the texts that surrounded its publication context were aimed at a male audience, and it would fit in with Urie's imagined readers better this way.

While the representation of gender is unstable in both the dissemination of these works and in the works themselves, the instability of science and art in Algarotti's works comes across as innovative rather than exclusionary. Algarotti's essay on opera merges scientific and artistic terminology in its discussions of eighteenth-century opera. During a time when disciplinary boundaries were still permeable, Algarotti's ability to draw on both the arts and the sciences enhances his writing. Patricia Howard describes Algarotti's work on opera as "central to the debate" about how opera should be produced in the mid-eighteenth century, because Algarotti "addresses opera both as a philosophical conundrum and as a living theatrical experience," and "his advice on theatre construction from the point of view of design, acoustics, and sight lines is still valid" scientifically.[123] Algarotti's introduction describes opera as a complex but still well-constructed "machine" and provides a mechanistic worldview of opera that aligns with mechanistic science.[124] Algarotti writes that "in forming [opera] no article was forgotten, no means omitted, no ingredient left unemployed, that could in any shape contribute to so important an end."[125] Echoing the mathematical teleology of the *Principia,* he thus adapts the connection between machinery and theatre—while not a new connection—for a Newtonian perspective. The work characterizes the ideal quality of opera, in English, as "pristine dignity."

The pleasing harmony of art and science also appears in *Newtonianism for Ladies,* where Algarotti writes that "the marvellous, of which the heart always desir[ou]s of being affected is so fond, happily arises in true philosophy of itself, without the help of machines," implying that the universe presents a pristine spectacle that an opera might emulate.[126] Algarotti's prose style in *Newtonianism for Ladies* was inspired by dialogue for a live performance—following the inspiration of Fontenelle's—"what I believed most proper for dialogue, clear, concise, interrupted, or interspersed with images and turns of wit."[127] He avoids phrases "which only serve to run the reader out of breath," implying that the work might be performed aloud if such a circumstance might arise, and giving a nod to contemporary reading practice, which often included reading aloud.[128] The theatrical comparisons are also discussed in the introduction: "I have endeavored as much as possible to render it lively, and make my readers interest themselves in it as they would in a composition for the Theatre. Is there any thing, especially where ladies are concerned, in which a writer should omit any endeavors to move the heart?"[129] Algarotti's readers would thus be captivated by drama, dialogue, and performance in printed form.

In addition to imagining a mechanistic perfection for opera, resembling the mathematical perfection of the Newtonian universe, Algarotti defines its negative, constructing poor opera as poor science. Opera is "liable to frequent failures and being out of order," as if it were poorly engineered.[130] He echoes the rhetoric of scientific experiment when discussing the good intentions of a failed opera: "So ingenious a composition . . . [may] not always answer the end proposed, although no care, pains or study, to attain perfection, had been omitted by those, to whom the preparation of such an entertainment was intrusted."[131] One might easily substitute the word "experiment" for "entertainment" and reframe this as a comment on experimental philosophy, the theatricality of which has attracted much recent inquiry.[132] These phrases also call to mind the marquise's sympathy for "philosophic worthies" in *Newtonianism for Ladies.*

According to this connection between experimental culture and entertainment culture, even if an opera does not work, Algarotti values the professional expertise of those who attempt to mount a production; perhaps this was why he expressed his fondness for Metastasio's libretto for Achilles on Skyros (also referenced in the essay on painting), an opera that had been produced many times—with uneven results—in Italy in the mid-eighteenth century. As sympathetic as Algarotti may be to well-intentioned failures, he still finds that opera has declined in quality. Using terms often associated with bad science, Algarotti claims opera has "dwindled into a languid, badly connected, improbable, grotesque, and monstrous aggregate."[133] In contrast to the perfect machine representing the ideal, contemporary opera is inert, desystematized, and disproportionate. It lacks penetrability, like a scientific text that speaks to no one. *Newtonianism for Ladies* describes outdated science similarly as "a chaos of vain and useless disputes [and] a chain of unintelligible definitions."[134] An earlier example described Descartes's discussion of globules as a kind of failed seduction, which parallels this association with vanity and lack of utility. Algarotti nonetheless describes opera as having a "friendly manner," something that had helped him connect with a broad audience in *Il Newtonianismo per le Dame*.

The concerns about opera's potential negative side effects parallel discussions about early novels from the mid-eighteenth century. Much as the words that describe poor-quality opera evoke bad scientific practices—"badly connected, improbable, grotesque"—novels that lacked credibility, verisimilitude, and narrative cohesion were likewise exposed to potential derision. Beyond the well-documented *Pamela* controversy in the 1740s, the public reception of novels as works that could potentially be educational or corrupt, depending on the hand that produced them, gestures to a broader problem. Creative thinkers in this post-Newtonian universe faced nearly unreachable standards of perfection—in terms of their ability to achieve complete narrative, operatic, or artistic cohesion, believability, and persuasion. Even eighteenth-century science stood largely awed by Newton; superimposing the standards of Newtonian science onto art thus produced an awkward set of criteria for works to fulfill. Even so, as in the case of bad science being compared to bad art, knowledge of standards for judgment was culturally necessary, especially to an emerging middle class.

Knowledge of the visual arts was another essential component of this cultural capital, and here, too, good art and good science align. In addition to the section on the painting of Achilles, Minerva, and their mingling globules, Algarotti discusses art elsewhere in *Newtonianism for Ladies*, including the examples of Raphael and Guercino, two artists who "had nearly an equal knowledge of the preparations necessary for the right designing of a figure, and yet were extremely unequal in the execution of it."[135] Algarotti again compares the occasional misalignment of artistic and scientific knowledge with the end product. The section on anatomy from the *Essay on Painting* details the kind of medical research necessary to paint successfully from life, stating that painters who have studied anatomy know "the better to understand the general effect, and remember the number, situation, and play of the muscles, it will be proper to compare, now and then, the anatomical casts, and even the dead body itself, with the living-body covered with its fat and skin."[136] Much as the *Essay on Opera* had used terms of good science to define good opera, this work aligns good-quality painting with scientific precision, including the Greeks as examples of anatomical accuracy because they were in better physical shape than the Romans. Not only were the Greeks scientifically and mathematically influential, but their bodies were better calibrated. This comparison inverts Algarotti's slight to Descartes, who in *Newtonianism for Ladies* resembles an ineffective lover because his globules of light dissolve after light inquiry. Readers of Algarotti's popularizations thus learned to judge inadequacies in opera, art, science, and sex.

Rebecca Messbarger's work on Anna Morandi Manzolini's connections between the fine arts and the medical arts also refers to the anatomical section of the essay on painting, so there were women who pursued this anatomical and artistic accuracy successfully.[137] Algarotti also includes popular tools like the camera obscura in the essay on painting as well as mentioning it in *Il Newtonianismo per le dame* in comparison to the eye, when he writes, "The eye is now discovered to have an entire resemblance to the artificial *camera obscura*."[138]

Algarotti appears in other works of the late eighteenth century, including *The Critical Review, or Annals of Literature* from 1773, the year after Newbery sold *The Lady's Philosophy* in his shop.[139] He is eulogized thus: "Count Algarotti died as he had lived, in the arms of learning, polite arts, and friendship."[140] *The Critical Review* uses terms that gesture to erotic, intellectual, and sociable registers—a combination that would not sound out of place in *Newtonianism for Ladies* or in Algarotti's other works on aesthetics. This may be a reference to the pleasurable seduction of *Newtonianism for Ladies,* even more so than to the works on painting or opera. The essay states that Algarotti's "well known dialogues on Sir Isaac's Optics were first conceived at Bologna, then composed at Rome and at Paris, often since corrected and improved in nine successive editions."[141] These lines imply that the dialogues were themselves gender-neutral, an implication corrected in the succeeding lines, which reveal that "the marchioness du Châtelet was very desirous not only of having them dedicated to her, but also of appearing in them as one of the principal interlocutors" in what is described as "Newtonianism for the fair sex."[142] The work was, of course, dedicated to

Fontenelle instead, who was still living at the time of its initial publication.[143] Du Châtelet remained present, silently, in the volume's original engraved frontispiece, where an Algarottiesque teacher lectures a beautiful female aristocrat modeled on her. Her anonymity here resembles Carter's exclusion from the title page of her own translation. Newtonianism, according to the print history of this work, was something for, rather than by, ladies, and the names foregrounded in print were male names.

When critics publish on Algarotti, they usually consider his cultural influence, his persona, and his correspondence. Historians of science use his work to make points about the far-flung nature of popular Newtonianism, or the republic of letters more generally. The interpenetration of art and science in the works I have described in this chapter, however, shows that Algarotti's role in the popularization of science and arts in English readership has been undervalued—all three of these works were published in multiple editions across the century, and from them we can learn much about the dissemination of popular science and its gradual integration with other emerging disciplines. Considering Algarotti and science as solely embodied by *Newtonianism for Ladies* fails to communicate the full extent to which Algarotti's writings as a corpus helped integrate science and the arts for an emerging, and diversifying, English-language readership.

This readership would grow increasingly diverse as more books circulated later in the century. My next chapter details the extensive information available about the New York Society Library and the five hundred readers whose borrowing habits from 1789 to 1792 reveal the exceptional diversity of popular science reading in the late eighteenth century. The influence of Newtonianism as a brand of knowledge still remained, even if Newton's direct influence was more diffuse than before. This emerging reading public nonetheless demonstrates the attractions of British science at a distance, and its reception in the new republic. These reading records also reveal links between readership and editorial and printing practices.

5 Reading Popular Newtonianism at the New York Society Library, 1789–1792

As we have followed the trajectory of popular Newtonianism in print, we have moved outward from the composition and reception of a single book—the *Principia*—through popular representations of Newtonian natural philosophy in different genres and at different times in the eighteenth century. I will conclude by analyzing popular Newtonianism from within a community of readers—this time, readers of diverse backgrounds who had the luxury of sampling the wide variety of books containing Newtonian natural philosophy that were on offer at the New York Society Library (hereafter NYSL) at century's end. These readers' consumption patterns reflect Newton's cultural significance and the global reach of British science, even in places as distant as eighteenth-century New York.

At the end of the eighteenth century, readers still avidly consumed works that explained Newtonian natural philosophy. For many, these explanations and this popularization of Newtonian ideas were cloaked in general information about scientific concepts. New Yorkers' reading patterns reveal the absence of a divide between expert and amateur readers, as well as further evidence that print practices adapted similar information about Newton to different audiences. But when does popular Newtonianism end in the print market? By the 1780s and 1790s, Newtonianism had been assimilated into print media such that amateurs and experts alike in New York encountered Newtonian natural philosophy across a range of texts. Reading about Newton no longer required picking up a book that mentioned Newton in its title as *Newtonianism for Ladies* had done; rather, works that taught astronomy, optics, and general knowledge proliferated and included Newton without advertising him. This shift does not mean that Newtonian popularization ended as a project, but it does reflect Newtonianism's assimilation into print. My focus on Newtonian science in an early American library follows Susan Scott Parrish's work on early American science, which analyzes in several rich chapters the connections between London and American science, English bodies' suitability to American climates, and the importance of correspondence between the Americas and the Royal Society, showing that the study of British science gains much from including New World locations.[1]

New York was an important place in the British (and early republic) Atlantic where Newtonianism was popularized during the eighteenth century. James Delbourgo's research on early American electricity has shown the intersections of British science and early American experimental culture. Early America responded to British and Continental science through its own experimentation. Experimental science relied heavily on Newtonian natural philosophy, as Delbourgo shows, especially at Harvard and in Philadelphia. Although Delbourgo describes British texts as influential, he spends less time on the printed dissemination of Newtonian natural philosophy and focuses less on New York, so my study will extend the work Delbourgo has done to cast our view of Enlightenment more broadly by focusing on readers of printed texts.[2] These readers' experiences can be traced through the charging ledgers of the NYSL.[3] Charging records like these, which rarely survive—and are rarely from the eighteenth century—offer records of multiple encounters with a single title or volume, and of borrowers' encounters with a title over time. A single copy of a title may be purchased once by a library, and then circulate zero times or a hundred times. Furthermore, a single volume of a multivolume work, such as the works of Buffon, may be far more popular than another volume, which allows us to understand better how Buffon's corpus was received among readers. Records for the borrowing of individual library volumes such as these offer exceptional opportunities for studying reception.

Unlike most libraries, the NYSL has detailed records for a wide cross-section of readers who had access to a broad range of scientific and popular science texts. It is an important library to scholars—a chapter on the NYSL opened Tom Glynn's recent study of the history of libraries in New York.[4] Its charging ledgers are especially significant because their survival is so rare: the library itself (est. 1754) still exists as a private library in New

York, and the ledgers have survived along with it. While comparable institutions exist that were founded in the eighteenth century, including the Liverpool Athenaeum and Boston Athenaeum, these institutions are both of later origin than the NYSL. The Boston Athenaeum has ledgers dating from the 1820s, a full generation after the period of my study. Those institutions that do have extant ledgers with eighteenth-century data include universities such as Columbia and Harvard, whose readers do not reflect as broad a segment of society as the NYSL, whose members included a watchmaker, several women, politicians, professors, and merchants. Popular books with scientific content, scientific research, and popularizations of science were read by many of the same readers and families at the NYSL, which shows that there was substantial overlap in the reception of these genres and extensive—and surprising—dissemination of scientific information within families and across networks. Circulating libraries of the eighteenth century—including larger libraries like William Lane's circulating library in London—catered mostly to the tastes of fiction readers published by presses like the affiliated Minerva Press, which Lane owned. Elsewhere in London, the British Museum's library was accessible to an elite membership or to those who were conducting research. When the rarity of the NYSL's ledgers is combined with the uniqueness of volume-borrowing data present in the ledgers and the fact that this particular library has a large number of borrowers with different backgrounds for whom biographical information survives, the NYSL is an ideal subject for studying the reception of scientific ideas in late eighteenth-century libraries.[5]

Beyond the unique suitability of the NYSL records for a study of popular Newtonianism, my focus on the NYSL follows the book historical tradition of exploring one institution in-depth, perhaps best demonstrated by James Raven's monograph devoted to the Charleston Library Society, which describes the NYSL as a peer institution.[6] Frank Felsenstein and James J. Connolly focus on a single institution for which extensive borrowing records survive from the late nineteenth and early twentieth centuries: the Muncie (Indiana) Public Library.[7] Patrick Spedding and Jennifer Furlong have used the NYSL ledgers in recent articles—Furlong focused exclusively on the NYSL, and Spedding compared it to a smaller library where borrowers were reading Eliza Haywood.[8] The NYSL, with nearly 500 members, has a healthy membership, similar to several institutions described in David Allan's work on subscription libraries, including the Liverpool Athenaeum (502 members) and the Norwich public library (456 members).[9] No one has described the interest in scientific reading that characterizes the readers at the NYSL. The NYSL's readers had a broad interest in scientific works, including the *Encyclopaedia Britannica,* which contains the most Newtonian content of the library's popular works and is also underrepresented in scholarship. As such, it will be discussed extensively throughout this chapter.[10]

Charging Ledgers and Popular Newtonianism

Readers did not need to encounter all of a scientific volume that attracted their interest in order to learn from it. Like Newton's *Principia,* where the front matter and introductory material were of exceptional importance, the lower-numbered volumes, and beginning sections of those volumes, remain essential when discussing library borrowers of scientific works across subgenres, and we can trace these reading habits to form valuable conclusions about which works attracted and held readers' interest.

One borrower who was interested in both scientific works containing original research and popular works with scientific content was John Stevens, Continental Congress delegate and successful businessman. The scientific works Stevens borrowed included Joseph Priestley's *Experiments and Observations on Different Kinds of Air* (1774–1777), Richard Watson's *Chemical Essays* (1782), Antoine-François de Fourcroy's *Méthode de nomenclature chimique* (1787), and the *Philosophical Transactions of the Royal Society.*[11] The large number of scientific works Stevens borrowed reveals that he had an interest in staying current in scientific research.[12] Stevens also borrowed the first six volumes of the *Encyclopaedia Britannica* as he was reading these scientific works; in spring and summer of 1790, two volumes (vols. 3 and 4) of the *Encyclopaedia* were loaned in between the *Philosophical Transactions* and Watson's *Chemical Essays.*[13] Someone with such an extensive interest in scientific content may have borrowed the encyclopedia volumes as light reading to learn about another topic in which he had less expertise, yet a reader like Stevens would be unlikely to ignore the scientific content of the *Britannica* that was close to his interests, especially during a period when he was reading a lot of scientific works. Stevens, then, probably learned some scientific information from the *Britannica.* Discerning the extent to which these volumes addressed Newtonian principles thus helps us to understand the knowledge of Newtonian science that was accessible to readers like Stevens toward the end of the eighteenth century.

Not everyone who borrowed in-depth scientific works desired comprehensive knowledge of contemporary science. Gulian Verplanck represents a different kind of reader who borrowed both scientific works and "popularizations": Verplanck seemed to be chiefly interested in works with a wide audience as well as anthologies.[14] On the spectrum of scientific expertise among readers of scientific texts, Verplanck would be categorized as a learned enthusiast, diverging from the more expert reading of Stevens. Verplanck's tastes included many volumes of the *Annual Register*, "or a view of history, politics and literature, from 1758 to 1779, inclusive," as well as several travel memoirs.[15] He also borrowed James Cook's *A Voyage Towards the South Pole, and Around the World* (1777), William Hunter's anatomy lectures, the Comte de Buffon's *Histoire naturelle, générale, et particulière* (1749–88), and Johann Kaspar Lavater's *Physiognomie* (1775–78), so he was unafraid of works with some scientific content, although works that were intensely mathematical were not on his list of borrowed books. Like Stevens, Verplanck also read multiple volumes of the *Encyclopaedia Britannica*, in between copies of the *Annual Register* and travel writings. This suggests that the *Britannica* appealed to a variety of readers with diverse interests in the arts and sciences.

If one were to chart the generic interests of readers who read the *Encyclopaedia Britannica*, Stevens and Verplanck would be nearer to one another than to readers who borrowed fiction exclusively, and there might be some utility to this exercise. Not all readers who had scientific interests borrowed works with scientific content, though, so this kind of interpretation would be limited in its applications. Malachi Treat was a physician and a professor of medicine at Columbia, yet his borrowing records reveal very little scientific content and almost exclusively fictional works.[16] From the borrowing records, one might inaccurately assume that no one in Treat's home had an interest in scientific topics, but of course this would not have been the case. The extensive presence of nonscientific works implies only that the household's borrowing habits at the NYSL were not scientific in nature; the family's interest in the *Encyclopaedia Britannica* was most likely not linked to scientific curiosity.

Subscription libraries were useful for the development of expertise, including eighteenth-century scientific expertise. Rebecca Bowd's research on subscription libraries in eighteenth-century Leeds describes the utility of subscription libraries, which "provided useful knowledge or technical information to increasingly professionalized groups like medical practitioners, bankers, overseas traders, and manufacturers."[17] Stevens shared his interest in scientific material with another reader. Based on evidence from the charging ledger, Stevens and Garrett Abeel read the *Philosophical Transactions* together.[18] Both men were merchants—one specializing in ship commerce and the other, iron. Both Stevens and Abeel checked out identical volumes of the *Transactions* at the same time.[19] Abeel's borrowing habits also include novels, works of travel such as Cook's and Hakluyt's voyages, Franklin on electricity, and John Winthrop's Massachusetts Journal.[20] As one of the more well-rounded readers at the library, Abeel's interest in science aligns with a general interest in staying informed. If one had to put together a composite of the stereotypical gentlemanly eighteenth-century reader, Abeel would be a good candidate. Many readers used the library to build knowledge of scientific principles. Popular science as a sub-genre was a smaller component of leisure reading than fiction, travel narratives, and other kinds of reference were; nevertheless, the NYSL's ledger demonstrates the varied exposure to scientific ideas that New York readers had in the late eighteenth century.

Current research on contemporary information-seeking has implications for studies of eighteenth-century reading, especially inquiry-based topics like scientific reading.[21] When we look at a charging ledger like this one, from a three-year span, we see a list of works consulted by individual readers during the first three years following the library's reopening at the end of the Revolutionary War, the volumes they checked out, and how long they kept the books. We don't see which specific questions readers had about the information they sought, but, rather, where they felt they might find answers. Different exogenous factors (such as interest in a topic, family makeup, the use-value of a particular book) and endogenous factors (the books in circulation during this period of time) affect these reading lists and practices. These exogenous factors vary and they mean that scholars cannot make too many assumptions about what the society library was to all of its patrons. Considering endogenous factors can more accurately shape our understanding of NYSL readership.

Two frequently borrowed works at the library that connected to science were Buffon's *Histoire Naturelle* and James Cook's Journals, which engaged scientific topics of weather, navigation, natural history, and astronomy. These would not be primarily considered popularizations of science, although they contained scientific information, but they were circulated widely and are important to examine for this reason.[22] Buffon's works well exemplify legitimate scientific content that was accessible to a popular audience. Between 1789 and 1792, 53 unique readers borrowed the works of Buffon, who had died in 1788.[23] These borrowers included some who may have shared the volumes in a household or discussed the works with one another, including the Van Horne family, the Johnson family, and the Stevens family. Other readers beyond these family links may have shared discussion of the books as well. W. Samuel Johnson, the president of Columbia College, and John Kemp, a mathematics professor at Columbia, both borrowed volumes of Buffon during the same year.[24] This may not seem like much of a coincidence, but because this was during a time when there were only two faculty members in the College of Arts at Columbia, the university's academic community was rather small. The most popular volumes of Buffon by far are the first three, which make up just over half the total number of loans, and the first four volumes were loaned a total of one hundred times. This differs from the borrowing pattern of novels, in which more than half of readers last to the end. Earlier volumes contain more of Buffon's theoretical material and less detail about particular animals, so readers of these volumes read more about scientific principles from Buffon than from some "popularizations" of scientific principles by other writers.

Following are the three volumes of Buffon with the most loans, as well as their subjects of interest:

Volume 1: 26 loans, subject: De La Terre (the earth)
Volume 2: 29 loans, subject: Animals, Generation, Human Development
Volume 3: 27 loans, subject: Le Cabinet du Roi

The Cabinet du Roi was an advertised part of the *Histoire Naturelle,* so the popularity of that volume makes sense, as readers would have sought it out. The popularity of the first two volumes also makes sense, as those were the most popular volumes of most books at the NYSL. The *Britannica* and William Smellie's translation of Buffon were also popular in Belfast. Richard Sher mentions that "the members of the Linen Hall Library in Belfast directed their secretary to purchase" these works among others, including Hume's History of England.[25] Andrew Bell, who had done the engravings for the *Britannica,* also did the engravings for Smellie's edition of Buffon in anticipation of having an Irish reprint to potentially help recoup expenses, so the print histories of the two works were intertwined as well as linked to Ireland.[26]

Captain James Cook's Journals, which offer scientific interest even though they are not original scientific research, show an even more pronounced imbalance than Buffon's works in terms of which volumes readers encountered. There were 90 unique borrowers of Cook—more borrowers than consulted the most popular novel in the library—and volume 1 of Cook was checked out 97 times.[27] These numbers reveal that readers who sought information on scientific topics found much of it in the low-numbered volumes of titles. Individual readers, however, had varied reading tastes when it came to science, and some were still interested in reading older works, scientific popularizations, and even less accessible scientific texts.[28]

Popular Newtonianism in the *Encyclopaedia Britannica*, Second Edition

Newton's ideas were disseminated through one of the most popular works at the NYSL, the *Encyclopaedia Britannica*. The *Britannica* reflected eighteenth-century intellectual development in print centers outside of London; it has been described by Sher and other scholars as an important printed work of the Scottish Enlightenment. Fara writes, "As Newton became more famous, natural philosophers recognized the advantages of allying themselves with his name. Labeling oneself a Newtonian came to command instant respect, and by the middle of the century, English natural philosophy was ruled by a Newtonian orthodoxy that was hard to contradict."[29] This was also true of the Scottish Enlightenment, which was closely linked with Newtonian natural philosophy. Scholars like Mark Towsey have described the extensive reach of Scottish Enlightenment texts in Scotland; Allan has shown British readers' responses to the Scottish Enlightenment throughout the eighteenth and early nineteenth centuries. The *Encyclopaedia Britannica* was a Scottish work of such interest to readers that the first edition was pirated in London and elsewhere, and the second edition was expanded by thousands of pages. The "Great EB" remained in print for over two hundred years, and it endures in electronic editions today.[30]

The encyclopedia enhances my study and helps us to fill in the gaps in understanding the reception of printed scientific works, because the encyclopedia is simultaneously a popularization and *not* a popularization. Its contents were not produced by amateurs to explain the work of experts, but often by experts in their respective fields. Even so, it was not an academic journal and was meant to be accessible to its readers. The second edition is rarely described in scholarship, with the exception of a thoroughly researched essay detailing its print history and its edition by James Tytler.[31]

The first edition had included many articles that remained for the second edition, which was by all accounts produced using inexpensive labor. Tytler, an ersatz apothecary and *balloonomane* who went into debt buying print supplies, helped correct errors in the first edition and supervised its significant expansion from three volumes into ten. The first edition by William Smellie had much rich material as a foundation for improvement, but Tytler deserves credit as "an outstanding encyclopedist" regardless.[32]

Sher writes that "as editor of the first edition of the *Encyclopaedia Britannica*, Smellie drew heavily on the published works of other Scottish literary men, such as David Fordyce, (whose *Elements of Moral Philosophy*, no. 7, was adapted for the article 'Moral Philosophy') [and] the obstetrician William Smellie (whose *Treatise on Midwifery*, no. 22, was used for the article 'Midwifery')" among others. William Duncan and Sir James Steuart's respective works contributed to "Logic," "Commerce," and "Money."[33] Of most interest to my analysis is the extensive use of James Ferguson's research on astronomy, wherein *Astronomy Explained upon Sir Isaac Newton's Principles* (1756) "formed the gist of the article 'Astronomy,' and whose *Lectures on Select Subjects* no. 61, was the basis of several other articles."[34] What is striking about this content being available in the *Encyclopaedia Britannica* is that Ferguson's *Astronomy* was also borrowed nine times on its own in the NYSL charging ledger, by seven unique readers. It was not actually borrowed by any of the same households that borrowed the encyclopedia, so with absolutely no overlap, two groups of readers were unaware that they were gleaning similar expertise on astronomy. Ferguson's title explicitly described its Newtonian origins, so in comparison to the readers of his monograph, the readers of the encyclopedia may not have known exactly how Newtonian their education was. This implies that Newtonianism had fully saturated late eighteenth-century culture to such an extent that it overdetermined reception: people were following nearly identical learning paths without even knowing. Richard Yeo describes the popularity of encyclopedias more generally, as "products of the period . . . of exponential growth in all things scientific—people, journals, data, and discoveries."[35] To that end, the *Britannica* is a component of the overall growth in information about scientific topics that circulated in the print marketplace.

One reader of Ferguson's *Astronomy* had an especially interesting sequence of loans with implications for studies of gender and readership: James Watson, Jr. borrowed Alexander's *History of Women*, followed by *Newtonianism for Ladies*, followed by Ferguson's *Astronomy*. The next book he borrowed was Baxter's *Matho; or, The cosmotheoria puerilis,* a natural philosophy popularization from 1740. All of these works were borrowed between late October and December 1791, so one book was likely in mind when a subsequent book was borrowed.[36] What seems evident from the sequence of titles, which begin with two female oriented books that are succeeded by two less-gendered works, is that the two explicitly female-oriented texts stoked an interest in learning more. Therefore, additional volumes with gender-neutral addresses were borrowed shortly thereafter. Even if Newtonianism could be "for the ladies," instruction in Newtonian principles did not need to end at the conclusion of Algarotti's book, but could lead to further study in a less-gendered, likely male-oriented book.

Sometimes readers in the same family had access to the same scientific information while reading different books. Daniel Crommelin Verplanck (1762–1834) checked out Ferguson's *Astronomy* three times from January through March of 1791, as well as other works by Ferguson, but did not read the *Britannica,* which builds on Ferguson's work.[37] His uncle Gulian Verplanck (1751–99), borrowed the astronomy-focused second volume of the *Encyclopaedia Britannica* the previous summer, reading it from July 22 to August 9, 1790.[38] Even though the nephew, Daniel Verplanck, seems like a more serious reader of popular science works than Gulian because he had specifically checked out Ferguson's *Astronomy,* Gulian had checked out five different volumes of Buffon earlier in 1790, as well as William Hunter's anatomy lectures, so he was not resistant to scientific content in his reading. Both of these men read broadly and encountered different forms of information. The next reader of Ferguson's *Astronomy,* Samuel Jones, read very differently from Verplanck—a lot of imaginative literature—but still checked out Ferguson without reading the *Encyclopaedia Britannica,* as did John Cozine.[39] Valentine Peters read Ferguson after having devoured each volume of the early eighteenth-century picaresque novel *Gil Blas.*[40] Looking at the borrowing patterns of these readers reveals a lot of flexibility in the conclusions we can draw about their interest in scientific content in general, and Newtonianism in particular.

The *Encyclopaedia Britannica*'s second edition was an important reference work and should have been of historical interest for American readers, because even though its writing began during the American Revolution, it was completed after the war. Some of the earlier volumes speak of the United States as a colony, whereas the latter volumes refer to the United States of America, so the print history of the second edition parallels the war and its aftermath. In spite of the second edition's potential interest to early republic readers, only 1,500 sets were printed, so the second edition was very rare compared to the first and third editions, which were in print for much longer.

The changes made to the third edition of the *Britannica* help us to understand the reception of the second edition. The third edition was advertised in an eight-page leaflet in 1788.[41] The leaflet mentioned many improvements over the second edition in a six-point list, including both cosmetic and content-related improvements. For one, it would "be printed in a style more suitable than any of the former editions, to the taste and opulence of the British Nation: the paper will be superfine, and the types occasionally renewed before they contract a worn appearance."[42] Elsewhere, the publishers mention that "at the close of publication will be delivered, An elegant Frontispiece, the Dedication, Preface, and proper Title-pages for the different volumes."[43] From these lines, we might assume that the encyclopedia was aimed at a refined audience that wanted to make sure the encyclopedia's appearance matched or superseded other works in their libraries. Even as it was being marketed to a wealthy market, the third edition was also sold by subscription, "published in weekly numbers, at one shilling each," so it was accessible to a broad range of incomes, and customizable to different interests. A person might buy the music or art "numbers," or those for history or science, each of which was "forty pages of letter proofs, closely printed."[44] For several shillings, a fair amount of expertise could be gleaned; for many pounds, a comprehensive, imposing set of this "Scientific Dictionary."[45] This advertisement for the third edition shows that it was expected to appeal to a wide audience, and that owning it displayed one's status. The use of high-quality paper implies that it could stand up to frequent use at home or as part of a circulating library.

The encyclopedia was an important work elsewhere in early America. Toward the end of the century, Thomas Dobson, a Scottish emigrant living in Philadelphia, produced his own reprint of the third edition of the *Encyclopaedia Britannica*.[46] Sher writes that Dobson "raised American reprinting to a new level" with this edition, "which he altered and retitled to suit his own designs."[47] Dobson printed between 1,000 and 2,000 copies of this edition for subscribers who included George Washington, Thomas Jefferson, and Benjamin Franklin.[48] He was able to succeed based partly on price: "By charging five Pennsylvania dollars per volume in boards, Dobson barely matched the one guinea price of the Edinburgh edition. The surcharge that

Americans had to pay for books imported from Britain enabled him to beat the Philadelphia selling price of the Edinburgh edition by 15 percent."[49] The Scottish Enlightenment was heavily represented in that volume as it had been in each of the Scottish editions: "the third edition of the *Encyclopaedia Britannica* devoted a great deal of text to Scotland" and "the American edition reproduced all of it," even though Dobson had revised the entries relevant to America, to some degree customizing the edition for his readership.[50] Sher also mentions that New York printers Robert Hodge and Thomas Allen partnered with Samuel Campbell to publish several works beginning in 1789, the early part of the charging ledger's dates.[51] There were, then, developing American printing houses producing similar works to those kept in the NYSL, while the NYSL was ordering many of its books from England.

Because I am analyzing an exceptionally rare edition of this encyclopedia, I will survey some of its contents for readers' benefit.[52] Newton appears throughout the encyclopedia, and the *Britannica* works with him in surprising depth.[53] Other reference works, like Chambers's *Cyclopaedia* (1728, 1753), had envisioned themselves as vehicles for transmitting Newtonian natural philosophy, so some Newtonian content would have been expected in later encyclopedias.[54] Volume 1, the most-borrowed volume at the NYSL, included a section on algebra on pages 224–37 with a section on Newton, so even readers who were learning about less sophisticated mathematics would have had some knowledge of Newton in that most-borrowed volume.[55] In volume 2, the most significant passage related to Newton is the more than 100-page section on astronomy with which that volume begins: pages 747–850, discussed at length later in this chapter.[56] Any reader who even skimmed the beginning of this volume would have had some exposure to Newtonian natural philosophy. Attraction has its own short entry, on the same page as "attorney"—a reflection of the work's eclectic content.[57] This section credits Newton with research on attraction and mentions that contemporary studies of electricity may account for Newton's theories about ether and its role in attraction.[58]

A person studying the popularization of science in the late eighteenth century might reasonably overlook a work like the second edition, when so few copies of it had been published. However, 74 readers at the NYSL consulted the library's one copy, potentially sharing it with many more in their households and families, so just one copy of that encyclopedia had a

Table 1. Borrowing records, *Encyclopaedia Britannica*

Year	Volume no.	Letter and page range	No. of loans
1778	1	(A, pp. 1–747)	39
1778	2	(A–B, pp. 747–1546)	28
1778	3	(C, pp. 1547–2364)	20
1779	4	(D–F, pp. 2367[*sic*]-3155)	34
1780	5	(G–J, pp. 3159–4002)	23
1780	6	(K–L, pp. 4007–4877)	17
1781	7	(M–O, pp. 4881–5558)	19
1781	8	(O–P, pp. 5561–6347)	10
1782	9	(P–S, pp. 6351–8075)	12
1783	10	(S–Z, pp. 8059–9200).	11

Source: "*Encyclopaedia Britannica*," New York Society Library: City Readers, http://cityreaders.nysoclib.org/Detail/objects/3275.

Notes: Volume 10 includes a 200-page appendix from 1784.

Other listings for the *Britannica* at the NYSL that are not clear in the ledger include the following: "11" 7 loans: (most likely roman numeral II); "22" 1 loan; "23" 1 loan; No listing: 11 times; "5 vols" x 1.

long afterlife.[59] The borrowing records of the *Encyclopedia Britannica* in the NYSL's first charging ledger break down as shown in table 1.

The NYSL records a purchase of the third edition in 1804, so this second edition was only in circulation for a short period. Hugh Gaine, a bookseller and trustee of the NYSL, published a 24-page catalogue of books for sale in 1792 that does not list this encyclopedia for sale, so it may have been hard to come by outside the library. Gaine not only was a trustee but had also published the 1789 NYSL catalogue, so he would have known the encyclopedia was in the NYSL's inventory.

Tom Glynn writes regarding the NYSL that "as a public collection, its development and use trace shifts in attitudes toward the kinds of knowledge that were regarded as socially useful and the bases of authority for disseminating such knowledge."[60] By Raven's account, most booksellers were negotiating with agents in London, which was recorded in the history of the NYSL. The library also received books from donors.[61] Beyond books from London, the NYSL held many other works of the Scottish Enlightenment and many with Dublin imprints. Sher uses both the Scottish editions of Buffon and the *Encyclopaedia Britannica* as examples of Scottish works that were frequently published in Dublin. Towsey shows that Buffon was very popular even in rural Scottish communities like Innerpeffray, as were Ferguson's popularizations of Newtonian science and mathematics.[62] Sher writes that the "Dublin booksellers reprinted vast quantities of books by Scottish authors. . . . They were selective, and generally chose the best books (in both a critical and commercial sense) in the various genres of polite literature."[63] The selectivity of the Dublin printers parallels the selectivity of society libraries as contrasted to subscription libraries: they were focused on acquiring works that could educate and interest readers.

The encyclopedia's 74 unique borrowers place it near the NYSL's most popular novels, *Honoria Somerville* and *Female Stability*, in popularity, but it was read differently than works of fiction, which require more linear borrowing patterns so that a reader may follow a particular plot.[64] Both the encyclopedia and Buffon have a similar borrowing ratio: loans of approximately three volumes per borrower who took it home. This trend continues across the earlier volumes of the encyclopedia. The first five volumes make up two-thirds of the total loans. People often read early volumes of a work, then either move on to something else or, in the case of a work like this that would serve as a home reference, might consider ordering a copy. One can also infer from these records that there were people in the late eighteenth century who intended to read the entire *Encyclopaedia Britannica* from cover to cover as a means of attaining broad and thorough knowledge. The *Britannica*, with 9,200 two-columned pages set in a small typeface, was a difficult but not impossible text to master. However, this mastery was not a feat the members of the NYSL managed to accomplish. Malachi Treat came close, but the first volume was checked out when he began to borrow it, and he also missed volumes 4, 5, and 6.[65] Garret Abeel seemed to be on track to borrow the volumes in order, but when he came to the library to borrow volume 4, volume 3 was on loan.[66] The reader who had checked out volume 3 when Abeel came to the library was John Stevens, with whom Abeel read related issues of the *Philosophical Transactions*, so the two men may have discussed that volume.[67] Henry Haydock Jr, governor of New York Hospital, borrowed every volume except for volume 6: someone almost encountered the entire work.[68] Volumes representing letters A–O, more than half of the alphabet, made up 80 percent of loans. The rest of O–Z and the appendix constituted 40 loans, 20 percent of the total.[69] The latter volumes are compressed, with S–Z being contained in one volume. The trends in volume borrowing further amplify the importance of the early volumes of the encyclopedia.

The presence of current information in the encyclopedia is significant. A reader searching for scientific information in an encyclopedia may have found much more up-to-date information in this text than in a text with a title that explicitly mentioned science, given some of the older research circulating at the NYSL. A reader who borrowed older issues of the *Philosophical Transactions* might be seen publicly as a reader who had access to scientific expertise, yet the reader of the encyclopedia would know more about contemporary science. Scientific content was not the only part of the encyclopedia that is shown to credit its readers' intellect: many classical literary examples present in the encyclopedia were presented in untranslated Latin, implying that readers either knew Latin or had the resources to procure a translation. Doig et al. write that this edition was "sometimes hostile to Newtonian natural philosophy," but still "endorses its methodological concepts."[70] This represents a departure from earlier encyclopedias, such as Chambers's *Cyclopaedia* and Harris's *Lexicon Technicum*, both of which, as Larry Stewart and Richard Yeo have identified, aligned their explanation of scientific systems with successful dissemination of Newtonian natural philosophy.[71] By century's end, Newton would be part of the system that would be explained to readers of encyclopedias, though not its sole participant.

Biographical Representations of Newton in the *Encyclopaedia Britannica*, Second Edition

One obvious place to look for the transmission of popular Newtonianism, especially in an encyclopedia, is in Newton's biographical listing, where his life and works are explained to interested readers. The *Encyclopaedia Britannica* contains lengthy biographical information about Newton. The summary of an individual's life, condensed for such a popular work, should reveal the person's cultural significance. Newton's entry, one of over 1,000 in the encyclopedia, was "one of the longest, and the longest on a scientist," reflecting his importance in the British and Scottish Enlightenment.[72] I will first describe biographical entries for other figures as a point of comparison. The presence of biographical information in the encyclopedia fostered readerly understanding of scientific figures. A biography of Robert Boyle appears in volume 2.[73] The biographies of the scientific figures at times verge on the literary in description. Small details are included in Boyle's entry, down to the sartorial: "His constitution was so tender and delicate, that he had divers sorts of cloaks to put on when he went abroad, according to the temperature of the air, and in this he governed himself by his thermometer."[74] Beyond the fancy cloaks is a lengthy discussion of why Boyle never married, even though he had opportunities.[75] Euclid receives two paragraphs, adequate biographical coverage for a mathematician about whom not very much is known.[76] Newton's contemporary Robert Hooke receives just over a page, and an unflattering portrait.[77] "He made but a despicable figure as to his person, being short of stature, very crooked, pale, lean, and of a meagre aspect, with dark brown hair, very long, and hanging over his face uncut and lank. Suitable to his person, his temper was penurious, melancholy, mistrustful: and, though possessed of great philosophical knowledge, he had so much ambition, that he would be thought the only man who could invent or discover."[78] With the exception of the odd physiognomy, this resembles a description of Newton that his detractors (like Flamsteed after their falling out) would share.

The biographical entry under Newton mentions that his father had an income of £120 a year, which is unlikely; it claims that Newton's mother married Ayscough rather than Barnabas Smith (Ayscough was her maiden name) and that she had no other children besides Newton, an incorrect reference, because he had half-siblings through his mother.[79] As to Newton's origins as a mathematician, a bucolic anecdote illustrates his humility and his destiny. Newton's uncle "found him in a hay-loft at Grantham, working a mathematical problem," which evokes the contrast between country origins and intellectual achievements.[80] Newton is also described as having been inspired by "the falling of some apples from a tree," rather than a more sophisticated explanation of the apple-moon anecdote that occurs in the astronomy section.[81]

Even within this title there are inconsistent representations of Newton that would have varied depending on the volume borrowed. Newton was also profiled under Astronomy, and appeared several other times in the encyclopedia. The extensive presence of scientific information meant a corresponding necessity to gloss Newton's achievements repeatedly. The Newton biography entry also claims to cite Fontenelle and describes Newton's lack of interest in Euclid, which we know is incorrect from Harrison's records on Newton's library, and from Westfall's biographical account. The encyclopedia claims that "in learning mathematics, [Newton] did not study

Euclid, who seemed to him too plain and simple and unworthy of taking up his time."[82] Although Newton did not read Euclid extensively at first, he was ultimately examined in Euclid by Isaac Barrow before his *Annus Mirabilis*, so Euclid was a part of Newton's formative mathematical education. The encyclopedia's representation of Newton and Euclid has more in common with Thomson's poem and its infantilizing of classical mathematics than with what we now understand of Newton's biography. Following its portrayal of Newton as condescending, the entry's standards for behavior slip further: the encyclopedia's Newton is lauded for refraining from beating his dog Diamond after Diamond knocked over a candle and burnt a manuscript: "Without once striking the dog, he only rebuked him."[83] These references to Newton's personality stem from Newtonian metaphors grounded in the *Principia* where Newton is assumed to be dominant and separate from other men.[84] To assume that each of Newton's actions stemmed from a cool confidence in his own superiority—that Euclid was "simple and unworthy" and that Newton would respond without temper to a manuscript being destroyed—derives from these representations.

Most significantly, this biographical entry insists on aligning Newton and print—a further link between bibliography and biography. The entry mentions that his Cambridge lectures were printed in English in London in 1728 in octavo, then in Latin quarto in 1729, a fact that attests to the high demand for printed works by and about Newton following his death in 1727.[85] The biography details the print piracy of Newton's Chronology, so the conflicts surrounding Newton's publication were considered important enough to memorialize.[86] This entry describes the *Principia* as a material object, as "a work looked upon as the production of a celestial intelligence rather than of a man."[87] The print history of the *Principia* is thus chronicled and linked to Newton's own characteristics and the "celestial intelligence" of his mind.[88] Indeed, this Newton "was a great lover of peace" who "never talked either of himself or others," so he apparently wins at modesty too.[89] Further to the passage's treatment of morality, the version of Newton in this biographical entry was "firmly attached to the Church of England," like an object bound to another object rather than a person who had his own beliefs.[90]

Some of Newton's personal characteristics come from earlier biographies, including Thomas Birch's 1738 biography of Newton, published in *A General Dictionary, Historical and Critical*, volume 7.[91] According to Birch's and the *Encyclopaedia Britannica*'s accounts, Newton's health was noteworthy, with the rare "incontinence of urine" listed among few things that plagued him in life. "Sir Isaac was of a middling stature, and in the latter part of his life inclined to be fat. His countenance was pleasing, and at the same time venerable. He never made use of spectacles, and lost but one tooth during his whole life."[92] Compared to Boyle, whose non-marriage rates a lengthy meditation, Newton "never married, and perhaps he never had leisure to think of it."[93] Perhaps this, too, reflects his masculinity—this Newton is so manly that the feminine never encroaches on his process.

His biographical entry also defines terms common to Newtonian science, so someone who wanted to know about Newton's personal life would learn about Newtonian science.[94] The authors summarize the Newtonian system, including definitions of place, space, motion, and absolutes. Other principles of the *Principia* are included, including axioms such as the laws of motion, as well as some objections to these axioms.[95] The entry even cites Newton himself.[96] Book 2 of the *Principia* is being described when the entry abruptly ends, before book 3 appears—perhaps there was an error in planning the entry's length, and the entry was simply cut short rather than edited.[97] The third edition did not amend this decision to end the entry at book 2. This may have meant that the entry on Newton was assumed to need no changes between the second and third editions.

The *Opticks* receives comparatively less treatment in Newton's biographical entry than the *Principia*, although the encyclopedia states that the *Opticks* is "a piece of philosophy so new, that the science may be considered as entirely owing to our author."[98] This places the date for the popularization of the *Opticks* as a recent one; even though it was published in 1704, by mid-century and later, Newton's *Opticks* was still seen as fresh and innovative. The entry on optics that concludes the volume actually offers additional coverage on Newton. Someone reading the biography on Newton would have a different perspective on the *Opticks* than someone reading the section on optics would have. This section describes Newton's optics and the *experimentum crucis* used to demonstrate that light was composed of particles.[99] It also finds Euler and Newton compatible and describes the experiments of Dollond.[100] When Newton is found to be mistaken, the encyclopedia still describes him in flattering terms: "It is no reflection upon Sir Isaac Newton, who did so much, to say that he was mistaken in this particular case."[101] Newton remains a conquering hero in this entry, more so than others whose contributions are detailed. He rescues the telescope as one would a dying baby, having "saved this noble invention from well-nigh perishing in its infant state."[102] One difference between the entry on optics compared to the others that mention Newton is his description mostly as "Newton" here (for example on page 5,545), versus the friendlier "Sir Isaac" in the other passages.

Front Matter and Other Newtonian Content

Much as the *Principia*'s front matter had helped readers assimilate its content, the front matter of the *Encyclopaedia Britannica* previews its contents and supports its editorial priorities. The title page describes the *Britannica* as "a dictionary of arts, sciences, etc, on a plan entirely new: by which, the different sciences and arts are digested into the form of distinct treatises or systems, comprehending the history, theory, and practice, of each, according to the latest discoveries and improvements." The preface contends that "from the feeblest efforts of infancy to the last convulsive struggles of departing life, these grand objects, these irresistible attractions, actuate all his powers, and animate all his enterprises through every gradation of his progressive being," in a humanistic embodiment of the principles of the *Principia*. Humanity is figured as masculine; the mind is, provocatively, described as a "she." Elsewhere the encyclopedia discusses its stance on popularization, stating that "abstract truths have, as it were, been clothed with a body, that by the drapery of narrative and allegory, they might be more effectually recommended to our notice and more agreeably inculcated." This comparison, echoing the discussion of a robed Achilles in Algarotti's *Essay on Painting* and the stripping bare of nature in Newtonian poetry, aligns the presentation of information with clothing, yet here truth should be clothed in appropriate narrative in order to be understood, rather than be exposed like a naked body.[103]

Under his editorship of the second edition, Tytler supervised content related to Newton. In the section on algebra in volume 1, a printed marginal note directs a casual reader to "Sir Isaac Newton's theory of divisors."[104] According to the encyclopedia, "A very ingenious method was invented by Sir Isaac Newton from finding the divisors of the absolute number by which the value of the equation is expressed; each of these was to be substituted in place of the unknown quantity, till some of them was found to answer the terms of the question."[105] One place where a contemporary reader would expect to see Newton is in a discussion of calculus, especially more than seventy years after the calculus dispute with Leibniz, after Euler had helped to standardize calculus notation. Nevertheless, the literal definition of a stone or pebble is the first one noted in the entry "Calculus," and readers are instead referred to the algebra section in volume 1.[106] Actual calculus discussion appears later, under the term Newton used, "fluxions."[107] In general, when a topic can be discussed under an earlier letter of the alphabet, it is, as in the examples of algebra, arithmetic, chemistry, and astronomy, each of which is covered thoroughly.

Newton figures heavily in the section on astronomy. The second volume begins with "astronomical tables for calculating eclipses," followed by a definition of astronomy. Astronomy "is the knowledge of the nature and properties of the heavenly bodies; their magnitudes, distances, and motions, both real and apparent; together with the natural causes by which their revolutions are performed."[108] Near the beginning of the entry on astronomy—the first section of the second volume of the encyclopedia—the writers discuss Newton, while at the same time not teaching readers very much. "The merit of Sir Isaac Newton is too well known to need any encomium, and his discoveries too numerous to be mentioned in this place. It is sufficient to say, that the science is perhaps more indebted to him than to all the other astronomers that ever existed."[109] Indeed, a reader would learn more about Tycho Brahe or Ptolemy from the historical background on astronomy present in this volume than he or she would about Newton, but—significantly—the reader would be encouraged to place Newton at the top of the scientific order without knowing his work in depth. Section 1 of this history of astronomy relates to "the apparent motions, magnitudes, and changes in the celestial bodies, as seen by the naked eye," so unlike the *Principia*, the *Encyclopedia Britannica* begins with the objects of sense perception, and the extension to the naked eye that the telescope, a prosthetic for the eye, introduced.

The astronomy section yielded information about Newton that Doig et al. describe as connected to Tytler's desire to "unseat the Newtonian view of the universe."[110] He refers to Hugh Hamilton's theory of comets, attempting to use it to help refute Newton.[111] A casual reader flipping through the pages would still encounter a mini-narrative of Newton's work from the marginal notes: "Hamilton's refutation of Newton on comets," followed by "Sir Isaac's account defended" and "Dr. Hamilton's hypothesis insufficient."[112] Hugh Hamilton (1729–1805, pub. 1767) relied on experimental evidence from Boerhaave and Woodward in order to try and refute Newton's theory of comets. The inclusion of Hamilton a decade later shows that the Britannica was interested in bringing the most up-to-date research to readers. Doig et al. describe this section as one that continually "raises objections to the Newtonian system;" the constant return to Newton and his pervasiveness throughout the encyclopedia subdues many of these objections.[113]

Following this section, there is a relatively short history of astronomical discoveries, including the following sections, which orient the history of astronomy around Newton's discoveries:

75 Tychonic System
76 Inquiries concerning the causes of the planetary motions
77 Cartesian System
78 Gives way to the Newtonian
79 Foundation of Sir Isaac's System
80 His discovery of the law of gravity

Section 80 includes the apocryphal story about Newton, the moon, and the apple. Newton, according to the *Encyclopaedia,* looks at the apple, then at the moon, and speculates like a character in a story: "Why not as far as the moon? (said he to himself)."[114] Even as it contains actual summaries of current research, the section on astronomy still participates in the folksier aspects of Newtonian popularization. The encyclopedia portrays Newton using terms that link back to his use of mathematical extrapolation in the *Principia.* Newton "undertook the reformation of philosophy, [and] proceeded upon a very different plan from all those who had gone before him, as he proposed to assume nothing as an hypothesis which was not deduced from what is obvious to our eyes; and thus, by organizing from those things which are within our reach, he thought we might come to know with certainty, what must happen in the celestial regions, to which access is denied us."[115] The idea of action at a distance is thus explained alongside the apple anecdote.

In sections other than his biography, historical information about Newton was ubiquitous in the encyclopedia. The encyclopedia's perspective on Newton appears populist in this passage regarding John Hutchinson: "He was contemporary with Sir Isaac, and expressed no small inveteracy against him, but without success; his unintelligible manner of writing rendering him in a great measure inaccessible even to his friends, and his malevolence disgusting every body else."[116] When comparing this perceived accessibility to the kind of intellectual inaccessibility and unintelligibility with which others have characterized Newton's own writing in the *Principia,* the encyclopedia attributes inaccessibility to another writer besides Newton. In fact, Newton scholars have ascribed several of these characteristics—including grudge-bearing, using intentionally challenging forms of expression, and inaccessibility to friends—to Newton himself.

Each volume of the encyclopedia contains at least one reference to Newton, so everyone who checked out a volume could have learned something about Newton's works. In volume 6, where he is least represented, he is still in the section on mechanics. When Newton's ideas are challenged, it seems to be done reluctantly, although Doig et al. think there is more behind the challenge in Tytler's editorial practices, including "attacks on Newton's mathematics."[117] Yet a printed marginal note reveals that the "dispute concerning attraction [is] of little consequence," and the text elsewhere dismisses criticisms of Newtonian attraction.[118] Later in the entry, the trend continues, and even though the "Newtonian system [is] overturned by admitting the action of electricity in the heavens," any "conjecturers ought to be indulged with caution."[119] Newton remains the standard for knowledge throughout the encyclopedia, with more recent thinkers' contributions defined against Newton. Section 5 of the astronomy chapter is described as "the Copernican system particularly considered and explained, on the Newtonian principles," rather than simply "the Copernican system."[120]

The encyclopedia's treatment of Newton was not interpreted as wholly positive at the time of publication. John Rotheram, an early reviewer, noticed a resistance to Newtonianism in Tytler's edition and took offense, writing that "every objection against the Newtonian philosophy is here set forth in ostentatious parade . . . for the sake of swelling out an article."[121] A reader who wished to purchase a full set might thus give pause to investing in it, or he or she might infer that laying out contemporary refutations of Newton was an attempt to remain current in the updated edition. What one person thinks is "swelling out an article" with refutation, another might find useful for recommending further reading. Readers who borrowed an isolated volume at the NYSL years after publication may not have understood Tytler's criticisms of Newtonianism as Rotheram did; as I have shown above, the encyclopedia's criticisms are couched in terms that reorient arguments back to Newton or defuse them. Regardless of how the criticisms to Newtonianism were received, we can identify Tytler's editorial practices as helping to delineate a shift away from popularization of Newton himself and a shift toward the popularization of scientific ideas, with Newton as part of a constellation rather than its entirety.[122] This is not to say that Tytler was secretly a fan of Newton; rather, it was becoming more appropriate later in the century to display perspectives that contradicted Newtonian natural philosophy in popular works. Doig et al. explain this shift away from Newton as a movement toward the use of electricity to explain Newtonian principles.

Beyond Newtonianism, there was other significant scientific content in the encyclopedia; it was even possible to acquire applied science skills from reading the encyclopedia. Volume 7's section on midwifery, based on the writings of the obstetrician Smellie, offered astonishingly detailed images of pregnancy and childbirth, including what to do about unusual fetal presentations, from breech to transverse.[123] This section included illustrations of techniques that are used even today by midwives, including reaching a finger inside the fetus's mouth to encourage her to tuck the chin before birth. Detailed information about breastfeeding and supplemental feeding suggestions much like those used by midwives and lactation consultants today were also included in the encyclopedia. The section on medicine included a large section on miscarriage, and this, too, may have assisted a practitioner or amateur in need.

Doig et al. claim that the encyclopedia had a mixed reception in the late eighteenth century. They write that its reception was "less successful and prestigious than the third edition," and perhaps that is why many scholars have overlooked it.[124] Reviews, including Rotheram's, were mixed, yet the encyclopedia sold well, and legitimate works even cited its different editions. George Adams cited several articles from the *Encyclopaedia*. He cited "Animalcule" in his *Essays on the Microscope,* which the NYSL held.[125] The *Britannica*'s penetration among the readers at the NYSL included at least one work citing it, as well as related works by Ferguson and other writers whose texts were used as the foundation for encyclopedia articles. The *Britannica* circulated widely among the readers of early New York, with many of its volumes entering the homes of a cross-section of families. Its intersections with eighteenth-century reception history are, I have shown, as complex as its varied source material.

So what of the popularizations of scientific ideas that were published in the eighteenth century with the explicit intention to popularize science? How do these works fit into the reading record when readers are consulting volumes with other types of scientific content? The term "popularizations," when applied to works that explain complex concepts to a general audience, implies that print publication magically makes something popular, but this is, as we have seen, not something we can take for granted. Newton's ideas inspired popularizations that were published in several editions across the century, including Algarotti's *Newtonianism for Ladies,* from the 1730s through the 1770s. This title had fewer borrowers, at 15, than the others previously mentioned, and several were in the same families—Van Horne and Seaman—so the dissemination of this title may have been quite limited, similar to the circulation of a single later volume of some of the earlier titles. *Newtonianism for Ladies* was significantly more popular at the NYSL than Newton's *Principia,* which the library owned but which was not checked out during the three-year period of the 1789–92 ledger. Newton's prophesies were comparable in popularity, however, with 11 unique borrowers.

Readers of Newtonian popular science at the NYSL included Effingham Embree, the only listed borrower of another work of popular Newtonianism: Henry Pemberton's *View of Sir Isaac Newton's Natural Philosophy* (1728) during this three-year period. Embree had it for a month from February 21, 1791, to March 24, 1791.[126] His renewal of the volume on March 19 implies that some of it was read at some point, if only out of guilt for having kept it so long. Embree, a clock and watchmaker, checked out several other volumes that imply the scientific questions he was asking were practical ones, possibly related to his continuing education or to that of his four children.[127] When he borrowed Pemberton's book, more than sixty years after its initial publication, Embree had not borrowed a volume since the previous July, so it is doubtful that he went through a lot of scientific books, or had much time for extensive self-instruction. Pemberton's book was followed by Watts on education and then "Preceptor; or, a general course of education," which Embree renewed three times, keeping it out from March until June. In December, Embree borrowed *Bath Poems* and Hester Chapone's bestselling educational works.[128] These two works, checked out over the Christmas period, seem aimed at women. Mostly fiction followed, with the exception of the first volume of *Alexander's History of Women*—the same book that had set the Watson household on a path to Algarotti, and later to Ferguson's astronomy. It appears that the questions or answers Embree and family sought from his library books changed over time, beginning with more specifically scientific interests and ending—at least for the period documented in this ledger—with general education questions targeted to women in particular. Close examination of these borrowing records reveals how diverse the reading patterns of readers at the NYSL could be: even two readers who borrowed the same title were likely using it as a gateway to other works that might differ greatly.

Academic Reading at Columbia and the NYSL

Although I have focused on the NYSL, its readership can also be analyzed in relation to the other institutions it intersected, because the NYSL did not stand alone as a library in late eighteenth-century New York. Although extensive records for other institutions do not survive, several readers at the NYSL had accounts at both the library and Columbia College, two institutions that were long connected. Because Columbia's records survive in part, I can expand my discussion of readership further to include the habits of academic readers. Columbia's president William Samuel Johnson had donated books to the NYSL, and the college therefore had shares in the library. The library-borrowing records of John Kemp, a reader of predominantly scientific works at two different libraries, help us understand an academic reader's borrowing practices.[129] Kemp's borrowing habits show that the Society Library's collections strengthened (or disclosed the strength of) a small community of academic readers at Columbia.

Born and educated in Scotland, Kemp taught at Columbia and was a trustee of the NYSL from 1794 until his death in 1812.[130] He had a reputation as an especially strict and hardworking instructor, who was said to have conducted over 600 experiments a year. By 1801 Kemp and Peter Wilson, a classicist, were the entire arts faculty at Columbia; and only two additional arts faculty were hired from 1802 to 1817, five years after Kemp's death. At times, Kemp and Wilson had to vote on issues where three people were needed for a quorum. They were told simply to vote as if there were three people present.[131]

Kemp's borrowing habits from the 1789–92 charging ledger reveal that he used the Society Library much like an academic library. In addition to responsibility for his primary field of mathematics, Kemp also had to be knowledgeable in the natural, chemical, and physical sciences to do his job. Access to a good library was essential to his success, and the NYSL provided much of what he needed. He had thirty-four loans during 1789–92, compared to eight loans in 1793–94 at Columbia.[132] Most of Kemp's reading is professional, including a loan of volume 6 of Joseph Priestley's experiments, but no other volumes borrowed from Priestley at the NYSL during that period.

At Columbia, Kemp borrowed volume 1 of Joseph Priestley's works in July 1793, so with this title two different libraries functioned identically.

Even though he was a scientist with current knowledge in his field, Kemp borrowed works that connected to popularizations. He borrowed George Adams's *Essays on the Microscope* (1787), which had cited the *Encyclopaedia Britannica*'s second edition, twice at the NYSL.[133] Raven describes the popularity of this work among the members of the Charleston Library society, who also ordered reliable instruments from Adams, the instrument and globe maker to George III. Raven finds in Adams's book an "infectious enthusiasm" that "makes very clear the interest in acquiring instruments," rather than abstract knowledge of microscopy.[134] Raven's chapter on science in *London Booksellers and American Customers* focuses primarily on the interconnectedness of book buying, the acquisition of instruments, and the development of experimental culture in Charleston, as a "prehistory" to the interest in "scientific experiment and discussion at the associated College of Charleston."[135] It is one of few chapters in the book that neglects to mention the NYSL as a peer institution. The circulation of scientific books was in fact important at both institutions, and later in the century experimental science and instrumentation became essential to the culture at Columbia.

Kemp also borrowed popularizations like those of the Comte de Buffon, as well as Colin Maclaurin's *Treatise on Algebra* (1748) and Benjamin Robins's mathematical tracts (1761)—the latter of these multiple times.[136] Some of his titles, even in 1789–92, reveal that Newtonian natural philosophy remained strong among scholars and readers later in the century. Benjamin Robins helped to introduce Newtonian mechanics to military research in the early to mid-eighteenth century.[137] Colin Maclaurin was a brilliant mathematician in his own right who had been mentored by Newton, and who had published *An Account of Sir Isaac Newton's Philosophical Discoveries* (1748).[138] Kemp also read Desaguliers at Columbia's library and later a lot of Euler at the NYSL, so he did not take it easy on himself and tried to stay current.[139] Yet, as we have learned, the need to integrate Newton with any study of science was still necessary in the late eighteenth century. Most of Kemp's leisure reading at the NYSL implies educated tastes. In addition to Hume's history, he borrowed Johnson's poetry editions; at Columbia he borrowed works on musical acoustics and botanical gardens.[140] Neither library served an exclusively academic or entertainment purpose for Kemp.

We might assume that other readers of scientific texts at both libraries formed Kemp's reading network, but his network was more diverse than this. Peter Wilson the classicist, his only near colleague for several years, was a trustee of the NYSL as well, from 1798 to 1814, so the two men must have interacted regularly, and even their reading habits intersected.[141] Both Kemp and Wilson read Benjamin Robins's work at the Society Library as well as Samuel Johnson's edited poetry volumes, and Wilson returned Hume's *History of England*, volume 1, just two weeks before Kemp came in to the library to check out volume two. Even though someone else may have checked out volume 1 by the time Kemp came in, Kemp may have read Wilson's copy of volume 1 and therefore would not have asked for it. This theory is bolstered by the fact that Wilson, after borrowing volume 1, did not take out additional volumes of Hume, whereas Kemp borrowed additional volumes, so they may have shared Kemp's copies. Both men also borrowed James Bruce's *Travels to discover the source of the Nile* (1790). For his part, Wilson also shared reading habits with Johnson, Columbia's president. The two appear to have shared the works of Frederick II in April 1791. In the same month, Wilson checked out volumes 9, 11, and 13 of Frederick II, and Johnson borrowed volume 10.[142] Neither borrowed volume 12, even though it was in the library at the time. Kemp and Johnson also had some overlap, including both of them reading Buffon in 1790, but Kemp had returned his volumes six months before Johnson borrowed his.[143]

The connections between Newton and library history extend even further than his presence at the NYSL. Newton himself was linked to libraries in the Atlantic world. Jeremiah Dummer, the Connecticut colony's London agent, not only successfully persuaded Elihu Yale to make a large donation to set up a university in Connecticut, he helped to collect books for the university library at the same time. Dummer met with Newton as well as Halley, and the men sent copies of books they had written to be placed in the Yale library at its inception.[144] Some of the books William Samuel Johnson read before he was at Columbia, while he was still a student at Yale, may have been books that Newton and Halley donated.

Glynn describes the NYSL and Columbia as conservative institutions whose identities aligned, writing that "since its founding with the Society Library in 1754, King's College, now Columbia University, had remained steeped in the classics, a bastion of conservatism, elitism, and Anglicanism."[145] Glynn writes about the Columbia of 1829, when the University of the City of New York—eventually NYU—was founded. At least late in the eighteenth century, faculty borrowing records like these reveal that the college was composed of faculty who were trying to stay current in learning while managing extensive administrative and teaching responsibilities. The diverse faculty and administration could connect through reading. The NYSL fostered a community of readers at Columbia and provided faculty with essential resources while it built its own collections. As the preface to the popular *Encyclopaedia Britannica* reads, "Abstract truths have, as it were, been clothed with a body, that by the drapery of narrative and allegory, they might be more effectually recommended to our notice and more agreeably inculcated."[146] The NYSL's readers enjoyed both the clothed and naked truths of Newtonian natural philosophy as the century came to a close through popular works that addressed scientific topics, and these truths were packaged and draped in multiple, surprising covers.

University libraries and society libraries were linked to one another in cities like New York, as social institutions, universities, and cities grew together. Society libraries like the NYSL attempted to define themselves apart from the tastes of circulating libraries. They did so by focusing less on fiction, even though fiction was a popular draw at many libraries. But I do not think we can split our analysis to identify a "real" reader who liked novels versus an ideal society library reader who had wide-ranging tastes in socially privileged categories like political theory, history, science, and the classics. Rather, one reader's tastes intersected with the rest of his or her family who shared a library account, as well as with others who borrowed within academic, associative, and society library networks. At the turn of the nineteenth century, the reading life of academics like John Kemp intersected with professional and personal reading. During a time of social transition as New York recovered from the Revolutionary War, multiple library collections helped its academics stay informed in their fields.

When we ask what happened to popular Newtonianism by the end of the eighteenth century, we can see that late in the century, the information that library readers encountered about Newton would have included popular science, scientific journals, books that celebrated Newton, books that mentioned Newtonian concepts indirectly, and books that included Newton in an emerging history of science. These texts did not align with neat binaries of expert and amateur as has been previously assumed, and were not borrowed in predictable ways. Books were brought into diverse homes, and readers did not obey imagined rules about whether to borrow scientific works directed at amateurs, whether to read original scientific research, or whether to read works that, like the *Encyclopaedia Britannica*, included both transpositions of original research and explanations of complex concepts.

By the close of the eighteenth century, popular Newtonianism was varied, existed alongside popular literature, and was present in the households of both expert and inexpert readers.

Conclusion

One component of the late seventeenth-century print market, the *Philosophical Transactions of the Royal Society*, had offered a consistent influx of diverse titles under one masthead around which other scientific titles circulated in England and abroad. As with fictional works, scientific publishing expanded and diversified by the end of the eighteenth century. By the 1790s, the *Philosophical Transactions* was still being published, as one text in a broad field of scientific and mathematical works that ranged from Lavoisier's and Euler's professional publications to children's books, scientific periodicals, and imaginative works such as Algarotti's. This diversity is reflected in how the works were received: by century's end, families were encountering the same scientific content in different books, perhaps without even knowing it. The differences among editions of *Newtonianism for Ladies* are important to the study of popular scientific books, works in translation, and works used for self-instruction, but they are also important within the broader narrative of eighteenth-century print dissemination. *Newtonianism for Ladies* was a book that, like the *Principia*, was significant enough culturally to be published in multiple editions. Also like the *Principia*, Algarotti's work contained representations of Newton as remote and distant from ordinary people. This remoteness and distance, I have shown, stem from Newton's cultural persona, which developed from the *Principia*'s representation of Newton as an author inextricable from his book.

The preceding chapters have shown how Newton and print intersected over time, both during Newton's life and after his death. Texts that were by and about Newton, and works that explained Newton's ideas, all track back to the persona we first saw constructed in the *Principia*. I have shown how much we can gain by analyzing the different points at which Newton, Newtonianism, and print intersected following the *Principia*'s publication: all of these connections show how important print was to Newton's popularization. Further, the study of *Newtonianism for Ladies* and the *Encyclopaedia Britannica* show how smaller case studies summon our reevaluation of larger cultures of print.

Beyond its applicability to researchers of print and science, this book's argument has further implications for literature and literary study: the challenges Newton faced in mathematical method and rhetoric when writing the *Principia* parallel the literary problems that faced prose fiction writers of the eighteenth century. These common problems include the representability of the natural world and of human consciousness. Even today, scholars explore the extent to which theories of human motivation in psychology, concepts that are linked to the way we understand fictional characterizations, remain indebted to Newtonianism. Current representations of matter and the mind have moved beyond Newtonianism, but psychologists' understandings of motivation have remained largely Newtonian. What Gantt and Williams rightly describe as a problem in contemporary psychology, I see as an opportunity to forge stronger understandings of character and motivation in the literature of the eighteenth century. Motivation, then as now, implies a triangulation of momentum, representation, and interactivity that acts to shape characters as if they were objects in the Newtonian universe.

Building on Gantt and Williams's ascription of mechanistic determinism to human motivation, the slippage between theories of attraction in natural philosophy and the representability of romantic and sexual desire in imaginative literature deserves further study. The use of mathematical perturbation problems in the *Principia* is one example of how we may link Newtonian methods to efforts at literary verisimilitude: the way we represent reality need not be linear, and can often be a dialogic process where reality and imagination yield fiction. This process might, in the sciences, lead to the completion of a proof or achieving equilibrium; in literature, perturbation offers a matrix for the generation of characters, conflicts, and settings. Although Gantt and Williams criticize human psychology's use of mechanism to "open up a mechanical terrain essentially devoid of human intention, activity, and meaning," novels help us understand that this bleak terrain may move beyond its Newtonian origins so that it can be lush with prose.[147]

By the end of the eighteenth century, Newton himself had become a kind of fiction, a metaphor for the heights of intellectual accomplishment, along with that accomplishment's myriad implications—from an association with divinity to an association with an almost superhuman level of knowledge. To track Newton's print afterlife even further, one sees how Newton ends up relating to media circulation today. Newton's legacy today is that of someone who represented seemingly limitless intellect during his time: an intellect that no longer seems possible for people in our era of specialization.

Culturally, Newton is seen to possess the kind of infinite knowledge many people falsely associate with search engines and databases. The founders of Apple Inc. invoked Newton in their earliest graphic designs, which included Newton sitting under an apple tree, and honored the scientist again by naming their 1990s PDA product the Newton.

Poets like James Thomson had imagined Newton sitting in the sky, watching "his" universe move around, and he has ended in a similar location in the digital era. The same Newton who flew through space in poetry continues to achieve that apotheosis in virtual space, where he guides researchers invisibly through the vastness of digital archives: in Eighteenth Century Collections Online (ECCO), he sits atop the website's masthead as a representative of the comprehensiveness of ECCO's holdings, placed in a position of power above and between Benjamin Franklin and Samuel Johnson. Newton's name was used for the entire library catalogue of Cambridge University through December 2017. Google has celebrated Newton's birthday with a ceremonial apple tumbling down the white space of the virtual page.[148] Repositories of knowledge affiliate themselves with Newton, much as Algarotti elevated his reputation through his associations with the deceased Newton. The preceding pages have argued for the relevance of Newton to histories of print and culture while remaining informed by the history of eighteenth-century science. The connections between Newton, print, and readership have illuminated the richly complex dissemination of Newtonian ideas through print. The answer to the question "What happened to popular Newtonianism?" may be that it is still happening today, and that the book you have just read has helped further that project.

Notes

Introduction

1. Newton's early publications on optics in the *Philosophical Transactions of the Royal Society* were essential to his reputation among his peers as a researcher, but in terms of his reputation as a figure defined by print publishing, the *Principia* is the landmark publication in Newton's career.

2. These elements are presented and discussed alongside images from the *Principia* in chapter 2.

3. Mordechai Feingold's *The Newtonian Moment* and Milo Keynes's *The Iconography of Sir Isaac Newton to 1800* illustrate the extent of Newton's popular reach in the eighteenth century. Margaret C. Jacob and B. J. T. Dobbs's *Newton and the Culture of Newtonianism* is an excellent overview of the history of eighteenth-century Newtonianism.

4. Because this is a book about print, I do not focus on live demonstrations, but Al Coppola's recent book *The Theater of Experiment: Staging Natural Philosophy in Eighteenth-Century Britain* analyzes extensively the links between spectacle and science. Recently there has been increased attention to reevaluating Newtonianism in light of more current historiography. James E. Force and Sarah Hutton's edited collection *Newton and Newtonianism: New Studies* further exemplifies the need to reevaluate Newtonianism and what it means for the study of science and culture in the eighteenth century. Some popular works link Newtonian science to the industrial revolution and modern engineering, including Margaret C. Jacob and Larry Stewart's *Practical Matter: Newton's Science in the Service of Industry and Empire, 1687–1851*. Earlier books like Larry Stewart's *The Rise of Public Science: Rhetoric, Technology, and Natural Philosophy in Newtonian Britain, 1660–1750* describe the links between popularizers of Newtonianism and a public who was interested in funding or otherwise supporting their projects. My focus on the use of the print medium as a vehicle for disseminating Newtonian natural philosophy extends these conversations by using contemporary methods in book history and bibliography in order to reframe the relationship between eighteenth-century print media and Newton's ideas.

5. Fara, *Making of Genius*, 3–8.

6. Ibid., 20.

7. I use the term "community" in the book, although this book is influenced by Bruno Latour and others who have written about network theory. Wenger et al. show

some differences between communities and networks that do not align with Latour's definition but which help illuminate how I am approaching community:

> The *network* aspect refers to the set of relationships, personal interactions, and connections among participants who have personal reasons to connect. It is viewed as a set of nodes and links with affordances for learning, such as information flows, helpful linkages, joint problem solving, and knowledge creation.
>
> The *community* aspect refers to the development of a shared identity around a topic or set of challenges. It represents a collective intention—however tacit and distributed—to steward a domain of knowledge and to sustain learning about it. ("Promoting and Assessing Value Creation," 9)

8. The intersection of print practices and reception history in eighteenth-century science that I explore in this book has been comparatively undervalued in scholarship, even though scholars of literary history have interrogated the connections between print and readership. There has been much scholarship on the Royal Society and Newton, including works like Michael Hunter's *Establishing the New Science: The Experience of the Early Royal Society* and Marie Boas Hall's biography of Henry Oldenburg. Other scholars placed more firmly in the field of eighteenth-century studies have been interested in print and information during that period, often not in ways that connect to science. One exception is the collection *Scientific Authorship,* edited by Mario Biagioli and Peter Galison. Karen Gevirtz's work on women novelists and eighteenth-century scientific ideas regarding the self, in *Women, the Novel, and Natural Philosophy, 1660–1727,* probes associations between Newtonian science and literature.

9. Shapin and Schaffer, *Leviathan and the Air-Pump.* See also Shapin, *Social History of Truth*, and Haraway, *Modest_Witness.*

10. Chico, "Gimcrack's Legacy"; Coppola, "Retraining the Virtuoso's Gaze"; Shanahan, "Theatrical Space and Scientific Space"; Mannheimer, "Celestial Bodies."

11. The most significant recent work of this type is Jonathan Kramnick's *Actions and Objects from Hobbes to Richardson,* which links action to the extension of thought, taking the links between empiricist philosophy and eighteenth-century literature into the field of action. Courtney Weiss Smith and Helen Thompson both build on this new way of framing empiricism and eighteenth-century literature. Smith's *Empiricist Devotions: Science, Religion, and Poetry in Early Eighteenth-Century England* and Thompson's *Fictional Matter: Empiricism, Corpuscles, and the Novel* re-situate empiricism in the same conversation as science and literature, which foregrounds the importance of science and philosophy to eighteenth-century literary history.

12. Haraway, *Modest_Witness,* 2–3.

13. Ibid., 3.

14. Horne, "Satirical Letter," 2.

15. Coppola, *Theater of Experiment,* 88.

16. Horne, "Satirical Letter," 3.

17. Later works like Laplace's innovations in lunar theory are good examples of abstract Newtonian natural philosophy's successful applications. Laplace used calculus throughout his work—rather than the Euclidean geometry of the *Principia*—to advance Newtonian mechanics and open the field up for further research.

18. For more on the instability of print in the late seventeenth century, see Adrian Johns, *The Nature of the Book: Print and Knowledge in the Making.*

19. Although disciplines emerged in the nineteenth century, more recent scholars have looked to earlier works like the *Principia* to understand disciplines' origins. Robin Valenza's *Literature, Language, and the Rise of the Intellectual Disciplines in Britain, 1680–1820* offers a history of eighteenth-century science that starts to bifurcate following the *Principia* into two groups of amateurs and professionals. This book agrees largely with the trajectory of professional scientists that Valenza lays out.

20. The connections between the *Principia*'s publication and Newton's desire to set himself apart from the Royal Society's publication practices are well documented. See Westfall's *Never at Rest* and Johns's *Nature of the Book* for more analysis of the differences between Newton's publication and the *Transactions*.

21. I discuss some of the Royal Society–endorsed monographs in my chapter on the *Principia*.

22. I. Bernard Cohen's excellent introduction to his translation of the *Principia* (1999, reprint 2016) details Newton's methods, as does Westfall's biography. Guicciardini's *Isaac Newton on Mathematical Certainty and Method* presents new insights on Newton's methods as well as his philosophies, and his relationships to the works of Barrow, Wallis, and Descartes.

23. The term "paratext" includes anything in a book that exists outside the content of the text. Paratexts might include epigraphs, prefaces, typefaces, or illustrations. Gérard Genette's *Paratexts: Thresholds of Interpretation* presents an obvious starting point for my discussion of paratext in the *Principia*. Genette makes thorough investigations of the origins, functions, and types of paratext that surround published works.

24. Swenson, "Optics, Gender, and the Eighteenth-Century Gaze"; Girten, "Mingling with Matter"; Armintor, "Sexual Politics of Microscopy"; Guicciardini, "Role of Musical Analogies ."

25. Iliffe, "'Is he like other men?'"; Iliffe, "Butter for Parsnips."

26. Johns, *Nature of the Book*, 627.

27. For further information outlining the different steps in the creation, production, transmission, and reception of books, Robert Darnton's influential article from *Daedalus* presents helpful context ("What Is the History of Books?").

28. Powell, *Performing Authorship.*

29. Dry, *Newton Papers.*

30. Westfall, *Never at Rest;* Cohen, *Introduction to Newton's "Principia"*; Iliffe, "'Is he like other men?'"; Iliffe, "Butter for Parsnips"; Fara, *Making of Genius*; Valenza, *Literature, Language, and the Rise of the Intellectual Disciplines.*

31. Gelbart trans., *Conversations on the Plurality of Worlds;* Loveland, "French

Thought in William Smellie's Natural History"; Doig et al., "James Tytler's Edition"; Costa, "*Ladies' Diary*"; Rider, "Perspicuity and Neatness of Expression."

32. Guicciardini, *Reading the "Principia."*

33. Ibid., 6.

34. Ibid., 10–11.

35. Rosenblatt, *The Reader, the Text, the Poem;* Iser, *Act of Reading*; Chartier, *Forms and Meanings*; Chartier, "Laborers and Voyagers."

36. Chartier describes material texts as "objects and forms whose structures and modalities govern their reading or hearing, and consequently the possible comprehension of the text read or heard" ("Laborers and Voyagers," 50).

37. Iser's reference to the "blanks" that must be filled in during the reading experience is also important to remember here, insofar as reading scientific works by Newton involves both abstract and direct spatial concepts.

38. Agamben, *Potentialities,* 179.

39. Ibid., 180.

40. Elaine Scarry describes this process in *Dreaming by the Book,* and Andrew Piper later discusses it more materially in *Dreaming in Books*.

41. St. Clair, *Reading Nation in the Romantic Period.*

42. To move forward past the literal assimilation of content, I recognize the importance of the connections between human beings and their networks, as well as the networking of ideas that print facilitates. Bruno Latour's writings have thus influenced my work here (see *Reassembling the Social*). Reception history traces the function of different actors (actors may include readers, writers, translators, the mechanism of print, spaces like libraries, and the texts themselves) within the network of Newtonian dissemination.

43. Concepts related to Latour's actor-network theory have been in place for a while in works like Shapin and Schaffer's *Leviathan and the Air-Pump* and Shapin's *Social History of Truth*, but were framed in relation to the establishment of credibility rather than reception among readers.

44. Delbourgo, *Most Amazing Scene of Wonders;* Bannet, *Transatlantic Stories and the History of Reading*.

45. Parrish, *American Curiosity.*

46. Brewer, *Afterlife of Character.*

47. Ibid., 24.

48. Indeed, we are frequently reminded when reading accounts of Newton that he was human and not a character, whether it is Voltaire's confirmation that Newton was "like other men," or whether it is the circulation of the charming anecdote about Newton and the apple, which gives readers a clearly defined visual to understand Newton's mind.

49. Wenger et al., "Promoting and Assessing Value Creation," 9.

50. There has been much scholarship on the Royal Society and Newton, including works like Hunter's *Establishing the New Science* and Hall's *Henry Oldenburg*. Other

scholars placed more firmly in the field of eighteenth-century studies have been interested in print and information during that period, often not in ways that connect to science. One exception is the collection *Scientific Authorship*, edited by Biagioli and Galison. The narratives of print technology and scientific discourse have much to offer one another, as scholars of earlier periods, like Anthony Grafton and Elizabeth Eisenstein, have elsewhere described. Karen Gevirtz's work on women novelists and eighteenth-century scientific ideas regarding the self probes associations between science and literature (*Women, the Novel, and Natural Philosophy*).

51. See Boyle, *Swift as Nemesis*, on more links between eighteenth-century culture, Newtonian natural philosophy, and criticism of modernity. Boyle argues that "A Tale of a Tub" includes a critique of the foundations of Newton's *Principia*. Swift's knowledge of the *Principia* has been debated; nonetheless, its first English edition was published alongside *Gulliver's Travels*, so at least the Motte family would have had familiarity with both works.

52. Even analyses of the *Principia* by expert readers of its time were not always consistent. Guicciardini explains that "the diversity of judgments of Fontenelle, Whewell, and Marie are caused by exposing the *Principia* to different audiences, who approach it with different background knowledge and for different purposes" (*Reading the "Principia,"* 3).

1. The Weight of Authority in Newton's Reading and Correspondence

1. Buchwald and Feingold, *Newton and the Origin of Civilization.*

2. Anthony Grafton describes some of the limitations of early modern humanists' reading, in particular their attribution of heavily weighted authority to classical texts. Grafton asserts that humanists "had confused the 'antiquity' of the Greeks and Romans—the fact that their texts had existed for a long time—with the authority that human beings gain as they age—an authority that can be invested only in people, who continue to learn as they age, not in books, which are impervious to experience" (*Bring Out Your Dead*, 100).

3. Blair, *Too Much to Know.*

4. Grafton, *Bring Out Your Dead.*

5. Many scholars have focused productively on Newton's readings of particular kinds of text, such as the *Principia* (Guicciardini), or his interest in particular genres, such as alchemy (Dobbs, Newman) or chronology and history (Buchwald and Feingold).

6. Jürgen Habermas argues that texts such as periodicals reflected the ways that personal identity began to orient itself toward an implicitly public audience (*Structural Transformation*, 43).

7. Latour's work on individuals and networks in *Reassembling the Social* is useful to consider here; in this case books are important actors in the reading communities of late seventeenth-century Britain.

8. Jennifer Andersen and Elizabeth Sauer's edited volume *Books and Readers in Early*

Modern England: Material Studies showcases extensive analysis of the various material traces that readers left in early modern texts. William Sherman's essay "What Did Renaissance Readers Write in Their Books?" analyzes the diverse ways in which early modern readers wrote in the margins of books, concluding that practices of marginalia are more complexly motivated than it might seem. In Newton's case, so much of his reading evidence survives in the form of manuscripts and correspondence as well as marginalia that it is possible to draw some conclusions about reading, as I do in this chapter.

9. John Harrison's *The Library of Isaac Newton* presents a wealth of information on Newton's library. Because it is the most comprehensive study of Newton's library, Harrison's work has been essential to this chapter. However, Harrison's work was published before recent archival shifts such as the movement of the Burndy Library and Dibner Collection to the Huntington Library in San Marino, CA. There has also been thirty-five years' critical engagement in book history since Harrison's text was published. Expanding on the study of some books Newton owned to make claims about how he used and looked at them, as I do here, is an important step.

10. Newton even crafted shelves to house the books he inherited from his stepfather at his birthplace in Lincolnshire.

11. See *the Newton Project* for nearly all of these manuscripts in digital form. As of this writing, more than six million words of Newton's manuscripts have been transcribed, and most of those written in other languages have been translated (http://www.newtonproject.sussex.ac.uk/).

12. For more on the trajectory of Newton's manuscripts, see Dry, *Newton Papers.*

13. Ibid.

14. Harrison, *Library of Isaac Newton*, 147, 131–32.

15. Miller, "Masculinity, Space."

16. Ovid, *Metamorphoses* (1593), 90–91.

17. In her analysis of Newton's alchemical works, Betty Jo Teeter Dobbs describes Newton's use of the *Metamorphoses* for alchemical experiments, such as the fusion of Mars (iron) and Venus (copper) into a blue alloy called "the net" (*Foundations of Newton's Alchemy*, 161–64).

18. Ovid, *Ovid's "Metamorphosis" Englished* (1678), 49.

19. Ovid, *Metamorphoses* (1593), 90–91.

20. Willard, "Ovid and the Alchemists," 153.

21. See also Miller, "Masculinity, Space."

22. Dog-earing, in particular, presents challenges to approaching the materiality of Newton's reading because of the difficulty in understanding what was meant by the page being turned down in the first place. It is difficult to tell whether the recto or verso side was initially turned down, whether the dog-ear was meant to point to the text, or whether it was simply intended to be a place marker. Because dog-earing leaves so few traces of the reader, it is even difficult to date. Especially in Newton's Bible, which he owned and used for sixty years, it is impossible to tell at what point in his

life Newton marked a particular page and why. At one time he may have been reading while an undergraduate at Trinity College; at another time he may have been reading as a scholar attempting to prove complex theological points. Some small claims can be made regarding Newton's interest in what he dog-eared, but these claims should be taken with caution.

23. Newton and Flamsteed had been friends but clashed over the delayed publication of Flamsteed's lunar tables, which Newton needed for his own benefit. Flamsteed so disliked Newton that he abbreviated his initials as "S.I.N." in large letters; he even attempted to insult Newton publicly in a preface to his *Historia Coelestis:* "I had formerly tried [Newton's] temper and always found him insidious, ambitious & excessively covetous of praise & impatient of contradiction. I had taken notice of some faults in the fourth book of his *Principia* which instead of thanking me for, he resented ill yet was presumptuous of his Interest that he sometimes dared to ask why I did not hold my tongue" (Suppressed preface).

24. Newton, *Correspondence*, 4:84.

25. Ibid., 4:110.

26. Markley, *Fallen Languages*, 156.

27. Athanasius was the bishop of Alexandria in the fourth century AD and was later Pope Athanasius I and St. Athanasius. He debated the validity of Arsenius's theories regarding the Holy Trinity at the Council of Nicaea in 325 AD. Athanasius believed that the Holy Trinity was unified and consubstantiate.

28. Markley, *Fallen Languages*, 145.

29. Newton, "*Paradoxical Questions.*"

30. Ibid.

31. Ibid.

32. Ibid.

33. The fact that there were multiple copies implies that at least one circulated.

34. Newton, "*Paradoxical Questions.*"

35. Ibid., f12–13.

36. Yeo, *Notebooks, English Virtuosi.*

37. Newton, *Correspondence*, 4:405.

38. Newton's Bible, his Book of Common Prayer, and his Psalter were bound together in one volume. They were given to Trinity College, Cambridge's library by Joseph Cox. Cox's father-in-law, Edward Folding of Hurstborne, gave it to him. The Bible's earlier provenance is that it was given to the nurse who attended to Newton shortly before his death in 1727.

39. Deut. 22:13–18.

40. Deut. 16:16.

41. Newton, *Yahuda Ms. 41.*

42. Buchwald and Feingold, *Origin of Civilization*, 127–28.

43. For more on Newton's *Chronology*, see Buchwald and Feingold, *Origin of Civilization.*

44. Newton, "Seven Drafts of Newton's Defence," 5r.

45. Spacks, *Privacy*, 7.

46. Barker, *Tremulous Private Body*, 7.

47. Dear's article "Totius in Verba" is a cornerstone of this conversation; Shapin's article on solitude, "'The Mind is its own Place,'" looks at the figure of the philosopher as a solitary one. This solitude stood in tension with the sociability of the early Royal Society.

48. Newton, "A Letter," 3075–76.

49. Newton, "A Letter," 3075. Westfall, *Never at Rest*, 158–67.

50. Newton, "A Letter," 3076.

51. We will return to temporality at other places in this book, because it is constant in the popularization of Newton: his permanence contrasts to ephemerality; time is a factor in the *Principia*'s mathematical proofs; the contrast between Newton and classical authorities persists across his career and print afterlife; and later, the borrowing records of actual readers show tastes in scientific material across time.

52. Newton, manuscript dated May 16, 1666.

53. "In a manuscript composed May 16, 1666 about solving mathematical problems through motion, in seven complex propositions, the last rule of which describes eliciting increasing and decreasing speeds and increasing or decreasing quantities of the related equation" (Whiteside, *Mathematical Papers*, 640–41; translation my own).

54. Westfall, *Never at Rest*, 58.

55. Ibid., 351 n55.

56. Ibid., 352 n55.

57. Newton, "Account of the Systeme," ff. 39–42.

58. Westfall, *Never at Rest*, 243.

59. Also see Shapin, *Social History of Truth*, chapter 8, for an analysis of Hooke's social position as a technician in the Royal Society. For more on Hooke, see Jardine, *Curious Life of Robert Hooke*.

60. Newton, *Correspondence*, 1:416–17. Their feud remained ongoing. Newton's skepticism toward scientific practices founded solely upon experimentalism was reflected elsewhere in English culture. As a literary parallel, Aphra Behn's and Thomas Shadwell's criticisms of experimental culture in seventeenth-century drama (*The Emperor of the Moon*, *The Virtuoso*) resemble Newton's criticisms, especially in their criticism of truth standards and the corruptible figure of the gentleman witness.

61. Newton, *Correspondence*, 2:441–43.

62. Another letter Newton wrote to Hooke offers telling criticism of the Royal Society's—and, by extension, Hooke's—epistemological claims. Newton criticized the *Philosophical Transactions'* reportage of advances in knowledge, writing "there is nothing which I desire to avoid in matters of Philosophy more than contention, nor any kind of contention more than one in print . . . what's done before many witnesses is seldom without some further concern than that for truth." In these lines Newton implies that print publication is inherently a public display with questionable motives.

The links between success and print publication, even in the face of piracy and other problems associated with print, is essential to Newton's career as well as the Society's. Johns's claim that the *Principia* is a significant end-product also gestures to its importance as a *beginning*-product, not just in the Royal Society, but in British and transatlantic print culture more broadly defined.

63. E.g., Johns, *Nature of the Book;* Shapin and Schaffer, *Leviathan and the Air-Pump*; Shapin, *Social History of Truth.*

64. It is important to point out some contradictions here. Newton was very interested in sound methodological platforms, but as later president of the Royal Society, he had a great stake in successful experimentation. Francis Hauksbee, Newton's protegé and maker of instruments for the Royal Society, performed important experiments in electricity with Newton's full support. Following many years of experiment, Hauksbee eventually published his research.

65. The University of St. Andrews has an excellent site contrasting the *Transactions* in 1673, under Oldenburg, and under Newton in 1723.

66. *Philosophical Transactions* (1685), 835.

67. Ibid. This volume was edited by William Musgrave with Francis Aston. Halley would take over editorial work for the *Transactions* in 1686.

68. Newton, *Correspondence,* 2:441–43. These "smatterers" published monographs that were vetted by the Royal Society as well as the *Philosophical Transactions*. For example, two post-*Micrographia* texts like the *Principia* used their front matter to present themselves as affiliated with the Royal Society: John Gedde's book on bee-houses (1675) and Alexander Pitfeild's [*sic*] translation of Perrault's *Memoires pour servir à l'histoire naturelle des animaux* (1687–88). These books reveal the uniqueness of the *Principia* in the context of the late seventeenth-century monograph.

69. Iser's description of print as a blueprint is useful to remember here, as the interpretation of both a blueprint and a printed essay from the *Philosophical Transactions* attempt to construct (sometimes literally) what is represented in print. Elaine Scarry describes the instructional properties of literature in *Dreaming by the Book*, including the ability of a text to represent surfaces passing over surfaces in order to create the illusion of solidity in a text. To take Scarry's point further, the dimensionality of print creates its own solidity mirroring this, with a book's individual fragile leaves, its miniscule peaks and valleys of chain and paper sitting uniformly on top of one another in bound form, heavy, leather-covered, and solid as an animal.

70. Or, as Johns writes, "intellectual history cannot just be the history of intellectuals" (*Nature of the Book*, 56).

71. Ibid., 541. See also Johns's *Piracy* for more on the history of piracy.

72. For more on Perrault's project, and its materiality as a "paper zoo," see Guerrini, "'Virtual Menagerie.'" The 1688 printing of Pitfeild's translation that Guerrini cites, printed by T. Basset, also prints his name with the variant "ei" spelling.

73. An imprimatur is an authorization for a book to be published and demonstrates that the material has been vetted in terms of its accuracy and/or its moral content. The

latter designation was especially important in books published with the imprimatur of the Catholic Church.

74. Advertisement, *Philosophical Transactions* 16 (1686–87): 297.

75. Johns, *Nature of the Book*, 100, 320.

76. James William Johnson's biography *A Profane Wit: The Life of John Wilmot, Earl of Rochester* includes discussion of the play, contextualizing it among historical dramas featuring sodomy and interpreting its satire of contemporary court culture under Charles II.

77. Johns, *Nature of the Book*, 320.

78. Fara, *Making of Genius*, 17.

79. The first edition of the *Principia*, according to the statute, would have been under copyright for twenty-one years, which means it would have expired in 1708.

80. Newton, "Letter to Halley," Oct. 18, 1686.

81. "Notes on the editing," *The Newton Project*.

82. Powell, *Performing*.

83. Ibid., 3–4.

84. Spacks, *Privacy*, 7. Spacks writes that "the writer and the reader function simultaneously in privacy and as members of a community." Newton had a heightened awareness of all of these roles: his position in relation to communities; his reading; and his writing. Newton had a great need for what Spacks refers to as "psychological privacy," in contrast to physical privacy, which refers to the physical act of sequestering oneself. Although he had a roommate while at Cambridge, Newton's family's estate at Woolsthorpe where he experienced the *anni mirabili* of 1665–66 afforded Newton his own room with private bookshelves. Spacks writes, "The subject of privacy . . . often demands focus on the ways people expose and guard themselves in relation to limited numbers of others. Within the private life—the life of people operating in the family, or in relatively small communities of friends—many forces impinge on the privacy of individuals, their capacity to protect themselves from other people's desire to know about them or to insist on their participation in social activity" (4).

2. The Attractions of Print in the *Principia*

1. I focus on the three Latin editions of the *Principia* published during Newton's life (1687, 1713, 1726) because I am writing about the links between Newton's use of print and his reception. The *Principia* was not translated into English (1728) until after Newton's death (1727).

2. Iliffe, "'Is he like other men?'"; "Butter for Parsnips."

3. Iliffe, "Butter for Parsnips," 33–34.

4. Iliffe's essay provides an excellent contextualization of the *Principia* and its divergence from the empiricist writing style favored by Boyle and the Fellows of the Royal Society, as well as a more comprehensive publication history of the *Principia*. I

treat several of the same topics Iliffe does, but within the framework of print history, and with an eye to the *Principia*'s translatability to literary studies and other studies of popularization.

5. These assertions are supported by Niccolò Guicciardini in *Reading the "Principia,"* which describes the ways learned readers encountered and responded to Newton's text. Guicciardini explains that "the diversity of judgments of Fontenelle, Whewell, and Marie are caused by exposing the *Principia* to different audiences, who approach it with different background knowledge and for different purposes" (3).

6. Wilson, "Classical Poetry," 70.

7. Haraway, *Modest_Witness*, 24.

8. The importance of understanding the modest witness intensifies when we include other attention to eighteenth-century science and gender, especially Sandra G. Harding's description of the *Principia*—specifically Newton's laws of motion—as a "rape manual" for its glorification of a system meant to dominate and control a feminized Nature. To call the *Principia* a rape manual—or to do so for these reasons—ignores its complexity and Newton's historical context for writing it (Harding, *Science Question*, 113).

9. See Dear, "Totius in Verba," for more on the ironic importance of linguistic representation in the writings of the Royal Society in the late seventeenth century.

10. See chapters 6–8 of Cohen's *Introduction to Newton's "Principia"* for close analysis of the structure of books 1–3 of the *Principia.*

11. Bender, "Enlightenment Fiction"; Valenza, *Literature, Language.*

12. According to Descartes's theories, vortices, the plural of vortex, enabled planetary motion. The *Stanford Encyclopedia of Philosophy* describes Descartes's system of vortices succinctly: "The entire Cartesian plenum, consequently, is comprised of a network or series of separate, interlocking vortices. In our solar system, for example, the matter within the vortex has formed itself into a set of stratified bands, each lodging a planet, that circle the sun at varying speeds. The minute material particles that form the vortex bands consist of either the atom-sized, globules (secondary matter) or the 'indefinitely' small debris (primary matter) left over from the impact and fracture of the larger elements; tertiary matter, in contrast, comprises the large, macroscopic material element This three-part division of matter, along with the three laws of nature, are responsible for all cosmological phenomena in Descartes' system, including gravity" (Slowik, "Descartes' Physics," Pr III, 48–54).

13. The famous anecdote about Newton and the apple is not present in the *Principia*, although it has become a legendary trope for Newton's ideas regarding universal gravitation.

14. Euclidean geometry uses axioms, theorems, and proofs in order to establish geometric truths, such as determining that two triangles are identical given the equal measurements of certain of their angles and sides. In Newton's case, Euclidean geometry was chosen for the *Principia* instead of Cartesian geometry and, to a certain extent,

calculus. By using Euclid's established mathematical model for proof, Newton simultaneously embraced a wide audience and tied his research to the ancients' knowledge of mathematics.

15. I use Cohen and Whitman's standard English translation of the *Principia* in this book, except when discussing variations across different editions of Newton's text, in which instances the translations are my own. I specify which Latin editions of the *Principia* I use as I go. Cohen and Whitman trans., *Principia* (repr. 2016), 408.

16. Later works exemplify this shift; Defoe's *A Journal of the Plague Year* (1722) circulates among these absolutes and uses time, space, place, and motion to narrate an event—an unpredictable, incalculable plague—that would seem out of sync with a post-Newtonian worldview. Ilse Vickers's *Defoe and the New Sciences* describes Defoe's work as heavily informed by Baconian inquiry and the early Royal Society, especially when it comes to the establishment of credibility in Defoe's texts. While this argument is successful, I see some texts, most notably *Plague Year,* which is not the focus of her study, as informed by different kinds of motion and narration.

17. Cohen, "Guide," 11.

18. Newton, *Principia* (1687), 5.

19. Cohen, "Guide," 149.

20. Cohen, "Guide," 150.

21. Cohen and Whitman trans., *Principia* (repr. 2016), 416.

22. Newton, *Principia* (1687), 1; Cohen and Whitman trans. (repr. 2016), 403.

23. Newton, *Principia* (1687), 8. "When a place moves, whatever is placed in it moves along with it, and therefore a body moving away from a place that moves participates also in the motion of its place. Therefore, all motions away from places that move are only parts of whole and absolute motions." *Principia*, trans. Cohen and Whitman (2016), 403.

24. Newton, *Principia* (1687), 4, 5; Cohen and Whitman trans. (repr. 2016), 408.

25. Cohen, "Guide," 128.

26. This does not happen in one instance, but rather, in the scholium to prop. 93, book 1; in book 1, prop. 45; in the four examples from book 2, prop. 10; and in corollary 3 to book 2, prop. 10.

27. Guicciardini, *Reading*, 93.

28. Wilson, in ibid., 93.

29. This is similar to what Derrida describes as "the *esprit d'à-propos* . . . the art, the genius, but also the technique that consists in knowing how to grab an opportunity, to make the best of it, the best economy of contingency, and to make of the *Khairos* or the *Chaos* a significant, archivable, necessary, or even ineffaceable event" ("Typewriter Ribbon," 77).

30. The arbitrariness of the event, however, does call to question how much something like the *Principia* could qualify: "The eventness of the event requires, if one wants to think it, this insistence on the arbitrary, fortuitous, contingent, aleatory, unforeseeable. An event that one held to be necessary and thus programmed, foreseeable, and

so forth, would that be an event? But then this arbitrariness undoes the power and force of a performative, which, as I was suggesting earlier, tends always to neutralize the event it seems to produce" (Derrida, "Typewriter Ribbon,"158). It is also possible that, following this logic, an experiment is actually a forced event, and therefore not an event at all.

31. Courtney Weiss Smith describes the links between a mechanized version of the universe and the divine in her chapter on Newton in *Empiricist Devotions.*

32. Newton orchestrated the publication of the *Commercium Epistolicum* (1712) intended to document his priority over Leibniz in developing calculus.

33. Guicciardini, *Reading,* 108.

34. Cohen, "Guide," 123.

35. Guicciardini, *Reading,* 2.

36. Newton, *Principia* (1687), 250–51.

> *The moment of a generated quantity is equal to the moments of each of the generating roots multiplied continually by the exponents of the powers of those roots and by their coefficients.*
>
> "I call "generated" every quantity that is, without addition or subtraction, generated from any roots or terms: in arithmetic by multiplication, division, or extraction of roots; in geometry by the finding either of products and roots or of extreme and mean proportionals. Quantities of this sort are products, quotients, roots, rectangles, squares, cubes, square roots, cube roots, and the like. I here consider these quantities as indeterminate and variable, and increasing or decreasing as if by a continual motion or flux; and it is in their instantaneous increments or decrements that I mean by the word 'moments,' in such a way that increments are considered as added or positive moments, and decrements as subtracted or negative moments." (Cohen and Whitman trans., *Principia* [repr. 2016], 647)

37. Cohen and Whitman trans., *Principia* (repr. 2016), 647.

38. Ibid.

39. Regarding the conjecture that people were able to read Latin, even in the late eighteenth century, works were published that assumed at least some working knowledge of the language. In chapter 5, I examine the *Encyclopedia Britannica,* which in its explanations of literature included Latin passages without translation. Even readers of a text designed for common knowledge such as an encyclopedia were expected to either have some Latin background, or feel like they should. When contemporary critics immediately dismiss the *Principia* as inaccessible because it was written in Latin, it should be done with caution. Is the math difficult, perhaps impossible for many to grasp fully? Yes. Was the prose understandable? More than has been credited.

40. A lemma, from the Greek word for "received," is a proven statement often used to help a mathematician build from one theorem to another.

41. Cohen and Whitman trans., *Principia* (repr. 2016), 647.

42. Ibid., 392.

43. Ibid.

44. Cotes's use of the term "occult" in this passage is less extreme than in current definitions of the word, which carries associations with witchcraft and other pursuits such as alchemy. Occult, however, means unknown in this reference. Newton's occult reading included research on the atmosphere, on gravity, and even on alchemy. For more on Newton and alchemy, see Dobbs, *Foundations of Newton's Alchemy.* Newman's *Atoms and Alchemy* presents a broader history of the ways in which alchemy connected to scientific experimentation in the early modern period. Newman and Principe's *Alchemy Tried in the Fire* re-creates many of alchemist George Starkey's alchemical experiments. Keith Hutchison in "Occult Qualities" provides further context for the tensions between occult qualities and the increasing reliance on manifest and demonstrable qualities.

45. Guicciardini, *Reading,* 9.

46. Scanned images of each edition of the *Principia* are available on *Early English Books Online.*

47. Hooke, *Micrographia.*

48. Ibid., 1.

49. The "Definitiones" are followed by one additional horizontal line. Newton, *Principia* (1687), 1, 12.

50. Newton, *Principia* (1687), 13.

51. Bender and Marrinan, *Culture of Diagram,* 7.

52. Harrison, *Library,* 147; 131–32.

53. Descartes, *Principia Philosophiae* (1692) (and many other editions). I am by no means offering a comprehensive list of each book to which Newton responded, and Johannes Kepler's work (especially the *Epitomae Astronomiae Copernicae,* 1617–21) was essential to the *Principia.* Even so, it is important to recognize that he had choices and predecessors in monograph forms and rhetorical strategies. Shapin and Schaffer's *Leviathan and the Air-Pump* is an excellent counterpoint to what Newton was accomplishing in the *Principia*—Newton's return to metaphysics at a time of experiment's primacy. See Shapin, *Social History of Truth.*

54. Kepler did not distinguish his three laws of planetary motion from the other conclusions he drew; however, these laws—especially the law of elliptical orbits and the inverse-square law—were essential to Newton's work in the *Principia.*

55. Although this chapter focuses mainly on Newton's connection to scientific publishing, a companion chapter could be written that considers Newton's innovations from the standpoint of mathematical publishing, in particular the *Principia*'s connections to mathematical textbooks.

56. Simply put, the Ptolemaic system places the planets in revolution around the earth, and the Copernican places the earth and other planets in their orbits around the sun.

57. The dialogue format would also prove successful in popularizations of science,

including Fontenelle's *Entretiens sur la pluralité des mondes* (1687) and Algarotti's *Il Newtonianismo per le dame* (1737), both of which are discussed in chapter 4.

58. Galileo's choice of the dialogue as a form for his arguments gestures to figures of antiquity such as Plato and Aristotle, as well as to his own early modern court culture. Given that he participated in court culture as well as mathematics and science, it is unsurprising that Galileo used characters who interacted with one another via dialogue as a means of communicating his philosophy. Galileo's casting of Salviati to cite Galileo as a member of the academy also points to science as part of a complex web of community and social practices. Later in the seventeenth and eighteenth centuries, the dialogue form would be used for popularizations of science by Bernard Bovier de Fontenelle (1686) and Francesco Algarotti (1737) that contained female as well as male characters.

59. Galileo's *Starry Messenger* (*Siderius Nuncius*) helped elevate its writer because it celebrated the Medicis' power. For more on Galileo and court culture, see Biagioli, *Galileo, Courtier*.

60. Galileo had described inertia previously; however, making inertia the first Law of Motion was Newton's choice.

61. Descartes, *Principia*, book 2.

62. Descartes was, of course, highly skilled in mathematics; I am discussing how the book itself organized knowledge and inquiry to present them to readers.

63. Newton participated in what Peter Dear describes as a shift during the seventeenth century from philosophical narratives to mathematical ones (*Discipline and Experience*).

64. This poem is discussed in greater depth in chapter 3.

65. Newton even wrote to Henry Oldenburg, prior secretary of the Royal Society, that he wished to quit the Society because he never went to London from Cambridge and did not attend the Society's meetings. Oldenburg persuaded the reluctant Newton to remain in the Society. Newton eventually became its president.

66. Genette, *Paratexts*, 207.

67. Cohen and Whitman trans., *Principia* (repr. 2016), 380.

68. Barchas's work in *Graphic Design and the Eighteenth-Century Novel* illuminates the importance of paratextual features such as adverts, front matter, and illustrations to the reception of eighteenth-century works. My next chapter explores how this poem in particular helped to cement Newton's persona in poetry, as well as how the poem responds to both masculine ideals and community pressures.

69. Cohen and Whitman trans., *Principia* (repr. 2016), 382.

70. [Not paginated but would start on 531]. Translations mine.

71. Contemporary bias away from Latin as a common language, combined with current scholarship that privileges the intense analytical reading we perform as scholars, shapes our perceptions about what it was like to read the *Principia* at the time of its publication. Latin is found throughout eighteenth-century printed works, often untranslated (as we will see more of in chapter 5), and it was a near-universal part of a male reader's curriculum.

72. Cohen and Whitman trans., *Principia* (repr. 2016), 941.

73. Ibid., 386.

74. The ability to proceed securely through unknown terrain via the use of mathematics is one of the aspects of the *Principia* that most inspired writers and artists. See chapter 3 for more about the imaginative writing inspired by Newton's own work.

75. Cohen and Whitman trans., *Principia* (repr. 2016), 398.

76. Cohen, "Case of the Missing Author."

77. Newton, *Principia* (1687), 401.

78. Ibid., 402. Translations mine.

79. Ibid.

80. Newton, *Principia* (1713), 357, 359, 362.

81. Newton, *Principia* (1726), 389.

82. The Motte translation of the *Principia* from 1728 could also be categorized as a popularization of Newton.

83. Guicciardini, *Reading*, 170.

84. Fara, *Making of Genius*, 9.

85. Newton, *Correspondence*, 3:156.

86. Guicciardini, *Reading*, 175–76.

87. Iliffe, "Butter for Parsnips," 47.

88. Newton, *Correspondence*, 2:485–86.

89. Cohen and Whitman trans., *Principia* (repr. 2016), 470. Centripetal force is the force that propels an object's motion along a curved trajectory such as an orbit; a constant velocity refers to the object's unchanging speed.

90. Newton, *Correspondence*, 2:487–488.

91. Ibid., 2:489–90.

92. Ibid., 2:501.

93. Cohen and Whitman trans., *Principia* (repr. 2016), 383.

94. Coppola, *Theater of Experiment*, 136.

95. Newton's "General Scholium" to the second edition of the *Principia* (1713) makes explicit the work's connections to God. Newton's extensive interest in theology was reflected in his manuscript writings, as I discussed earlier in this chapter.

96. From *Politische Testament von 1752*, qtd. in Lualdi, *Sources of "The Making of the West*," 110–11.

97. I use the term "science" as an umbrella category to describe mathematical and natural philosophical pursuits during Newton's time, when disciplinary specificity is not crucial to the point under discussion. I use the terms "natural philosophy" and "mathematics" when they are necessary to point out disciplinary distinctions (for example, Newton's title at Trinity College, Cambridge was the Lucasian Professor of Mathematics, rather than "scientist" or "natural philosopher").

98. In addition to Genette's work with paratexts, this chapter was also influenced by Janine Barchas's work, which correctly asserts how much the look of books was an essential component of how they were read and whether they were bought in the first place. Barchas argues that "the eighteenth century novel . . . relies heavily on its own

physical presentation as a printed book to guide the reader in interpretation" (*Graphic Design*, 16). Similarly, I argue that the *Principia* deploys the form of the book to guide the reader in understanding and interpreting its points.

3. Newtonian Popularization and Masculinity in Two Poems about Newton

1. Fara, *Making of Genius*, 20.

2. Valenza, *Literature, Language*, 19.

3. Habermas, *Structural Transformation*. Lawrence Klein evaluates the relationship between gender and public and private spaces, rejecting the "domestic thesis" that binary gender divisions aligned with the binaries of public and private. The four contexts Klein presents for reinterpreting publicity and privacy include an "associative public sphere." Like the associative public sphere Klein describes, masculinity was also associative ("Gender and the Public/Private Distinction," 97).

4. Philip Carter, *Men and the Emergence of Polite Society.*

5. Shapin and Schaffer, *Leviathan and the Air-Pump.*

6. Wall, *Imprint of Gender.*

7. See the previous chapter for more coverage of the history of the *Principia*. Other works on masculinity during the early modern period abound in contrast to the eighteenth century, perhaps because of the homosocial nature of performance culture in the Renaissance and scholarly interest in the sexuality of James I. See Hammond, *Figuring Sex Between Men*, and Barnes, *Post-Closet Masculinities*.

8. Lanser, *Sexuality of History*, 2.

9. The aura attributed to Newton was reinforced by other early eighteenth-century representations of him as well. Richard Yeo posits that "even if readers consulted only one or two [encyclopedic] terms pertaining to natural philosophy, they would be drawn into a critique of Cartesianism and an affirmation of Newtonianism" (*Encyclopaedic Visions*, 166).

10. Parker, *Triumph of Augustan Poetics*, 62.

11. Ibid., 66.

12. Gabbard, "Clashing Masculinities."

13. Iliffe, "'Is he like other men?,'" 159. Recent articles such as Guicciardini's "Musical Analogies in Newton's Optical and Cosmological Work" offer appropriately in-depth research into the connections between Newton's writing and culture.

14. Powell, *Performing*, 2.

15. See Westfall, "*Principia*," in *Never at Rest*, 402–68; and also Iliffe, "'Is he like other men?.'"

16. Newton, *Correspondence*, 1:416.

17. See Johns, *Nature of the Book*, and Berry *Gender, Culture, and Society* for examples of the extent to which print was an unstable medium for the transmission of verifiable knowledge.

18. Feingold, "Mathematicians and Naturalists." Feingold writes that "the publication of the *Principia* marked a new phase in the relations between mathematicians and naturalists, not only because its immediate success contributed greatly to the widespread acquiescence to the claims of mathematicized physics . . . but also because the triumph of the Newtonian method appeared to sanction its application to other scientific domains" (87).

19. Westfall, "*Principia*," in *Never at Rest*. The Royal Society did not disapprove of the *Principia*; rather, they gave it their imprimatur and printed a prepublication review by Halley in the *Transactions*. However, Newton still worked outside the normal science practiced by the Society. Some Fellows, like Robert Hooke, felt they had not been given sufficient credit in the *Principia*.

20. Cohen and Whitman trans., *Principia (repr. 2016)*, 381.

21. Halley, "Accounts of Books," 291.

22. Ibid., 282–83, 283–90.

23. Ibid., 291.

24. Albury, "Halley's Ode." The poem was written in Latin originally and was not part of the series of popular Newtonian texts, fictional or otherwise, that were published after Newton's death in 1727. One exception to the sparse scholarly discussion of the ode is Albury's article from 1978 that reprints the entire Latin ode and a translation, in part because of the lack of a literal translation of the ode before that date.

25. Halley, "Ode," 52.

26. Smith, *Empiricist Devotions*, 73. Smith focuses on the role of religion in shaping representations of both mechanism and the occult force of gravity in popular Newtonian works in the form of "meditated empiricism," 92. Smith's work spends more time working with the popularizers of Newton than on their readership; her work makes the essential point that in studying popular Newtonianism, "a clean separation of public and private or science and magic is simply untenable" (76).

27. The Huntington's and the British Library's copies show that readers consulted the opening of the *Principia* heavily and left the rest unmarked. This is unsurprising since even Newton suggested that lay readers refrain from trying to read all of it.

28. As qtd. in Iliffe, "'Is he like other men?,'" 174.

29. As in the previous chapter, I use the standard Cohen and Whitman translation of the *Principia*.

30. Halley, "Ode," 31–32.

31. Ibid., 40.

32. Ibid., 33–34.

33. Ibid., 38–39, 41.

34. Ibid., 36–37.

35. Cowley, "To the Royal Society," 60.

36. Ibid., 99–100.

37. Halley, "Ode," 32.

38. Pepys, "February 1665." Sunday was often the day Pepys did accounts and

caught up on business. He spent "all the afternoon with W. Hewer at my office endorsing of papers there, my business having got before me much of late." According to Barker's *The Tremulous Private Body,* Sunday was also the day of the week when Pepys read the pornographic novel *L'eschole des filles*.

39. Halley, "Ode," 1.

40. Cowley, "Royal Society," 176–78.

41. Halley, "Ode," 2.

42. Ibid., 13.

43. Hooke also built a reflecting telescope in 1673.

44. Halley, "Ode," 28.

45. Shapin, *Social History of Truth,* 311.

46. Halley, "Ode," 22–24.

47. "Quaequae scholas frustrae rauco certamine vexant."

48. The power this poem attributes to mathematics anticipates Priestley's later comparison of elegant mathematical formulae to the sublime.

49. Kuhn argues that normal science offers explanations that account for anomalies, a process called "saving the phenomena." See his *The Structure of Scientific Revolutions.*

50. Halley, "Ode," 15–16.

51. Ibid., 18–20.

52. Philip Carter, *Men and the Emergence of a Polite Society.* Masculinity is essential to Halley's poem as well and is ubiquitous in eighteenth-century studies. Carter has expanded on connections between masculinity and sociability by contending that societal norms help to construct societal extremes; in Carter's argument, the fop's excessive sociability produces a deviant masculinity.

53. Cohen and Whitman trans., *Principia* (repr. 2016), 410.

54. Ibid., 412.

55. Halley, "Ode," 52.

56. Brewer, *Afterlife of Character,* 26.

57. Smith, *Empiricist Devotions,* 92.

58. Ibid., 91.

59. Parker evaluates Cowley as a poetic ancestor of Thomson, claiming that among Cowley's accomplishments was a certainty that "the language of science would replace the language of analogy" (*Triumph of Augustan Poetics,* 88). Thomson's ancestry is more diffuse than this, and includes writers like Halley, whose representations of Newton in the *Principia* influenced Thomson's poem about Newton. For Parker, "Thomson's nature is a place of natural beings and natural forces. His classicism is decorative and his Christianity is cold" (91). Yet the innovation of Thomson's time makes its poetry unique: "'Numbering the streaks of the tulip,' is in fact a marked tendency in Garth, Gay, Pope, Thomson, and Akenside, and their new descriptionist art is original to their own time" (123). Newton, whom Keats would later criticize for "unweaving the rainbow," also influenced Thomson, and presenting both literal representation and attuning to the artistic value of description is a feature of both Newton and Thomson.

60. St. Clair, *The Reading Nation*, 119.

61. Aikin, *Essay* (1777), 36.

62. St. Clair, *The Reading Nation*, covers this material, following Rose's thorough work in "The Author as Proprietor."

63. Qtd. in Rose, "Author as Proprietor," 54.

64. Thomson, "Isaac Newton," 1–4.

65. Ibid., 11.

66. Ibid., 2.

67. Desaguliers, *Newton*, 21.

68. Halley, "Ode," 18; Thomson, "Isaac Newton," 46.

69. Thomson, "Isaac Newton," 56.

70. Ibid., 13–16.

71. Halley, "Ode," 22–24.

72. Nicolson, *Newton Demands the Muse*.

73. Thomson, "Isaac Newton," 32.

74. Jonson, "William Shakespeare" (1623), line 39.

75. Thomson, "Isaac Newton," 35.

76. Ibid., 36–38.

77. Harding, *Science Question in Feminism*, 113.

78. Fara, *Making of Genius*, 23.

79. Thomson, "Isaac Newton," 23–29.

80. Ibid., 24.

81. Perhaps, to align this poem about Newton with the phenomena Brewer describes, the fictional works about Newton may simply be arguing that Newton is a character, forcing him into a discourse where he shares space with Gulliver, Macheath, Crusoe, and other popular fictional heroes of the period.

82. Thomson, "Isaac Newton," 57–58.

83. Ibid., 69–72.

84. Ibid., 76.

85. Ibid., 82.

86. Ibid., 90.

87. Halley, "Ode," 50.

88. Thomson, "Isaac Newton," 15–16.

89. Cohen and Whitman trans., *Principia* (repr. 2016), 940.

90. Thomson, "Isaac Newton," 39.

91. John Millan owned the copyright to three of *The Seasons*' poems; Millar owned the copyright to "Spring."

92. *Measuring Worth* gives its most conservative estimate for the 2015 value of £105 in 1730 as £14,370, and its largest estimate as £2,040,000. The value of £137 10s is £17,510–£2,655,000. Regardless of which figure we use, it represents a significant sum (*Measuring Worth*, June 15, 2016).

93. Thomson, "Receipt for Payment." When Andrew Millar purchased the rights

to these poems and essay, they were listed in the exact same order in the document assigning copyright to Millar, a direct quote from Thomson's receipt. Millar had already purchased the rights to "Spring" with the drama *Sophonisba* on Jan. 16, 1729.

94. Thomson, "Isaac Newton," 133–35.

95. Ibid., 144–45.

96. Ibid., 160.

97. Thomson, "Isaac Newton," 187–89.

98. Butler, "The Elephant in the Moon" (1784), 9–27: lines 1, 14–16. This edition preserves the standard version of the poem, its version in iambic pentameter, and a separate fragment of verse.

99. Butler, *Poetical Works, vol. 3*.

100. Butler, "Elephant in the Moon," 19–28.

101. Shank, *Newton Wars*, 398.

102. Swann, *Politics and the Parlement*, 99.

103. Fautrière is mentioned in Swann, *Politics and the Parlement*, 99.

104. Davy de la Fautrière, "Épître Newtonienne," 11–12.

105. Ibid., 73–85. "I ask Newton and this is his response: / 'In a given center, where all weighs the same / 'From a mutual attraction / 'Where all bodies prove laws / 'There is the universal cause / 'No living thing in the universe / 'Escapes these general laws / 'From the contact of central forces / 'Is born the variety of movement' / Content with such a clear lesson / My spirit tastes the sweetness / Of finally understanding the mystery / Of God the creator's works" (translation mine).

106. Ibid., 63.

107. Ibid., 86–91.

108. Thomas, *Eloge de René Descartes*, 92. Translation mine.

109. Ibid., 100.

110. Pope, *Dunciad*, 4.31–34.

111. Pope, "Eloisa to Abelard," 228.

112. Smart, "Temple of Dulness," 65–68.

113. Ibid., 75–80.

114. Ibid., 72, 70–71.

115. Hartman, "Confessional Counterpublics," 45–46.

116. The focus on "outness" is present in Newtonian scholarship as well, though there has been no consistent way to write about Newton's potential homosexuality, a subject to which scholars attribute varying degrees of weight and credibility.

117. Wordsworth, *Prelude*, 53–82; 63, 62.

118. Ginsberg, *Howl I*, verse 217.

119. Hartman, "Confessional Counterpublics," 51.

120. Ibid., 52.

121. Someone like Walt Whitman might represent a thematic transitional figure between Thomson and Ginsberg.

122. Ginsberg, *Howl I*, verse 218.

123. Ibid.

124. Reynolds, "'The Mad Ones,'" 158.

4. Extending Popularization in English-Language Editions of Francesco Algarotti's Works

1. Algarotti, *Opera* and *Painting*.

2. Other works of the period offered similar distillations of knowledge for autodidacts, such as reference works like dictionaries and encyclopedias, and books designed to teach foreign languages.

3. Some of this absence may be explained by the disciplinary positioning of Algarotti as a figure whose activities bordered on several disciplines. As such, scholars have tended to write about the texts of his that relate to their interests: opera historians write about his opera writings, etc. Those who have written about Algarotti in England have chiefly focused on the significant historical figure of his translator Elizabeth Carter, described more fully later in this chapter.

4. This is not new to Algarotti, insofar as Fontenelle was interested in gesturing to his audience's knowledge in the *Entretiens*, but Algarotti's ability to pivot and produce similarly layered, interdisciplinary texts on painting and opera is unique and underexplored.

5. There is extensive scholarship about public formation in eighteenth-century Britain, whether it builds on arguments in Anderson's *Imagined Communities* or engages with the ideas about bourgeois society in Habermas's *Structural Transformation of the Public Sphere*. Other scholarship builds on more recent works such as Warner's *Publics and Counterpublics* and Latour's *Reassembling the Social*. This chapter acknowledges the complexities of public life in England during the eighteenth century, including the links between print's modalities and an imagined reading public.

6. Guicciardini, *Reading the "Principia"*; Gingerich, *Book Nobody Read*.

7. Chartier, "Labourers and Voyagers," 56.

8. The English-language editions of *Il Newtonianismo per le dame* were published in 1739 (London: Cave), 1742 (London: Hawkins), 1765 (Glasgow: Urie), and 1772 (London: Newbery).

9. Other recent scholarship, including Jennifer Farooq's work on the history of early-to-mid-eighteenth-century sermon publishing, has gestured to the complexity of this market; Farooq finds that the "decentralisation of sermon publishing in the mid eighteenth century is more difficult to explain" than other parts of sermon publishing. Farooq's chapters on the printing and reception of eighteenth-century sermons help illuminate the diversity of the print marketplace in eighteenth-century London. Sermon publishers included Oswald, Ward, Noon, Buckland, C. Davis (who also printed Voltaire's *Letters on the English Nation*), and John Pemberton (Farooq, *Preaching*, 83). For more information on the reception history of sermons in particular, with her inclusion of primary source evidence, see Farooq, *Preaching*, 108–42.

10. Ibid., 83.

11. I also build upon Jeffrey R. Wigelsworth's analysis of early eighteenth-century editions of Desaguliers's translation of s'Gravesande's *Physices Elementales:* Wigelsworth focuses on the role of reputation and advertising in the popularization of Newtonian science across different editions of the same work ("Competing to Popularize Newtonian Philosophy").

12. See Barchas, *Graphic Design;* St. Clair, *Reading Nation*; and Raven, *London Booksellers and American Customers.*

13. The collection *Scientific Authorship,* edited by Biagioli and Galison, is one essential volume that fits this description. Johns's *The Nature of the Book*, an obvious influence on this project, likewise expands scholars' knowledge of the connections between scientific knowledge and publishing practices. This kind of interconnectedness among the philosophies and publishers of the Scottish Enlightenment is explored further in Sher, *Enlightenment and the Book* .

14. Chartier, "Labourers," 56. These texts further complicate the excessive and varied cultural dissemination of Newton following my analysis of poetry in chapter 3.

15. Voltaire, *Letters,* 117–18.

16. Fara, *Making of Genius,* 20.

17. Other works of Newtonian popularization from the eighteenth century include Henry Pemberton's *View of Sir Isaac Newton's Philosophy* (1728); s'Gravesande's *Mathematical Elements of Natural Philosophy, Confirm'd by Experiments: or, an Introduction to Sir Isaac Newton's Philosophy* (1747); *The Newtonian System of Philosophy, Adapted to the Capacity of Young Gentlemen and Ladies* (1761) by "Tom Telescope"; and Fontenelle's *Eloge de M. Neuton* (1728). Much important work remains to be done in connecting these printed works to a broader audience.

18. Hutton, "Women."

19. Agorni, "The Voice," 202 fn85.

20. In addition to the four extant English-language editions of *Newtonianism for Ladies* discussed here, Algarotti's works also made their way across the Atlantic; however, his books on painting and the opera are more popular in North American library catalogues of the eighteenth century than his work on Newton, which is seldom mentioned. An exception is the New York Society Library, which owned a copy of one of the English-language editions of *Newtonianism for Ladies,* in two volumes. The library's first charging ledger reveals that individual volumes of the book—referred to in the ledger as *Newtonian philosophy explained, for the use of the ladies, by Algarotti*—were checked out seventeen times between 1790 and 1792, but it is not clear which edition the library owned. All readers who checked out the book were male, but the NYSL website offers biographical information that indicates many were married or had families for whom the books may have been intended. For most of these readers, *Newtonian Philosophy* was the only work on Newton they ever borrowed from the library. See "Sir Isaac Newton's Philosophy."

21. Hutton, "Women," 189.

22. Findlen, "Translating the New Science," 203.

23. Valenza, *Literature, Language,* 81.

24. See chapter 3 for more analysis of these works and their connections to masculine print communities.

25. Mazzotti, "Newton for Ladies."

26. Shank, *Newton Wars,* 364.

27. Algarotti's bisexuality may also contribute to the diffuse definitions of gender in his texts, since he may have had awareness of multiple sexual roles.

28. Cohen, "French Conversation."

29. Fontenelle's influential and highly popular *Entretiens sur la pluralité des mondes* (1686) showcased intimate moonlit conversations regarding the Copernican and Cartesian systems between the text's male narrator and a marquise with a keen interest in natural philosophy. Fontenelle's work helped popularize Cartesian and Copernican science in Western Europe as Algarotti's text would later do for Newtonian natural philosophy, with a similarly intimate setting and format.

30. Hutton, "Before *Frankenstein,*" 21.

31. Moretti, *Distant Reading,* 206.

32. By contrast, the use of the indefinite article with regard to women often implies mysterious authorship or an unnamed dedicatee, as in "by a lady" or "to a lady." The Italian title did not even necessitate the inclusion of a definite article in the translation. *Newtonianism for Ladies,* as I have translated it here, is also a perfectly acceptable translation of *Il Newtonianismo per le dame.*

33. The readership I describe has increased substantially from the Latin readers of Newton's *Principia* (1687).

34. That is to say, male readers (Agorni, "The Voice," 192).

35. Hutton writes that Mary Wortley Montagu was one of the readers who believed Newton's calculations in the *Opticks* to be off-putting much like those of the *Principia*, but, as Hutton writes, "this did not mean that Newtonian science was beyond the comprehension of less gifted minds" ("Before *Frankenstein,*" 23).

36. In this chapter, I use Elizabeth Carter's 1739 translation of Algarotti's *Il Newtonianismo per le dame*, published as *Sir Isaac Newton's Philosophy Explain'd for the Use of the Ladies.* I also cite the original publication, *Il Newtonianismo per le dame* (1737), as well as the following translations: Duperron de Castera trans., *Le Newtonianisme pour les dames* (1738); *Sir Isaac Newton's Theory of Light and Colours* (1742); *The Philosophy of Sir Isaac Newton Explained* (1765); and *The Lady's Philosophy* (1772).

37. Carter trans., *Newtonianism* (1739), vii.

38. Arato, "Minerva and Venus."

39. Hutton, "Women," 186.

40. Carter trans., *Newtonianism* (1739), ix.

41. For more on Conduitt in Thomson's poem memorializing Newton, see chapter 2.

42. This contrasts to my discussions in chapter 3, where popular depictions of Newton resemble a superior man more than an actual deity.

43. Fara, *Making of Genius*, 66.

44. Valenza, *Literature, Language*, 83.

45. Carter trans., *Newtonianism* (1739), 203.

46. The repeated development of examples that relate to seduction and courtship echoes the repetition of proofs and diagrams in the *Principia*. See the end of chapter 2 for more on the narrative patterns of the *Principia*.

47. Carter trans., *Newtonianism* (1739), 108.

48. Ibid., 104–5.

49. Ibid., 110.

50. Ibid., 112.

51. Ibid., 113.

52. For readers interested in pursuing the figure of Carter more thoroughly, Mirella Agorni describes the circumstances surrounding Carter's role as translator of Algarotti in contrast to Behn's role in translating Fontenelle's *Entretiens*, the work from which Algarotti drew inspiration. Agorni does not focus on the subsequent editions of Algarotti as this chapter does. Fissell and Cooter explain the differences between Fontenelle's *Entretiens* and Algarotti's presentation of similar material in *Newtonianism for Ladies*, finding that the marquise in Algarotti's version, "already familiar with Cartesianism," is portrayed with a "less-artificial and less-condescending tone" (Fissell and Cooter, "Exploring Natural Knowledge," 137–38).

53. Agorni, "The Voice," 182.

54. There are several works on Carter for those with more interest in her biography and her place in the culture of eighteenth-century Britain. Among these essays are Agorni, "The Voice"; Hawley, "Elizabeth Carter"; McGeary, "Clarissa Harlowe's 'Ode to Wisdom'"; and Taylor, "Clarissa Harlowe, Mary Astell, and Elizabeth Carter."

55. Such periodicals as Eliza Haywood's *The Female Spectator* were aimed directly at women; Patricia Meyer Spacks's 1998 edition for Oxford University Press is a helpful anthology. Helen Berry's *Gender, Society, and Print Culture* analyzes a periodical that embraced both male and female readers. Shawn Lisa Maurer's *Proposing Men* delineates the relationship between men and eighteenth-century periodical literature.

56. The copies consulted are part of the Dibner Collection of the History of Science at the Huntington Library, San Marino, CA.

57. Moretti, *Distant Reading*, 204.

58. The NYSL's 1789–92 charging ledger enters *Newtonianism for Ladies* as "Newton Philos.," so libraries followed this gender-neutral titling trend as well.

59. Following are examples of footnotes that were common to both the 1739 and the 1742 editions:

1739, Volume 1, Dialogue 1

Page 2 in Dialogue 1, Guicciardini's History of the Wars of Pisa, 24.

Page 3 in Dia. 1, Fracastorius, 25.

Page 3 in Dia. 1, Streams of the Benaco, 25.

Page 5 in Dia. 1, Alphonsus the 10th King of Aragon, 27.

Page 6 in Dia. 1, Mandragora, 28.

Page 6 in Dia. 1, Aelius Sincerus Sannazarius, 28.

Page 10, Laura Maria Katherina Bassi, 32.

1742 Hawkins, Volume 1, Entertainment 1

Page 3, Guicciardini, 5.

Page 3, Jerome Fracastorius, 5.

Page 4, Lake Benaco, 6.

Page 6, Alphonsus, the 10th King of Aragon, 8.

Page 6, Mandragora, 8.

Page 7, Aelius Sincerus Sannazarius, 9.

Page 11, Laura Maria Katherina Bassi, 13.

60. Piper, "Footnote Detection." Piper breaks down his process for the digital/visual recognition and analysis of footnotes, but the items he considers important computationally build on Grafton's work and extend it to the kind of material questions I have been considering in this book. Of particular note is the contrast between the data I have found (in which footnotes decrease from the beginning of a text to the end, and from one edition to another) and his assertion that footnotes should increase over time. Piper writes, "*Footnotes per page.* If the thesis is that Enlightenment = indexicality, then we should find more footnotes as time progresses in our data set. The first, most basic question is, how many footnotes are there in this data and can we get a per page measure?" Of course, this chapter does not do data mining and as such cannot be seen as a direct competitor to Piper's research. But it is interesting to see the extent to which my conclusions align partly with Piper and partly with Grafton.

61. Duperron de Castera trans., *Le Newtonianisme*, 3.

62. Arato, "Intorno al *Newtonianismo*," 565–66.

63. Duperron de Castera trans., *Le Newtonianisme*, 16.

64. Shank, *Newton Wars*, 362.

65. For an excellent and thorough recent assessment of the state of inquiry into the history of intellectual property in book history, see McGill, "Copyright and Intellectual Property."

66. Guile, "Newtonian Pictures," 210.

67. Algarotti, *Il Newtonianismo per le dame* (1737), 1.

68. Carter trans., *Newtonianism* (1739), 1.

69. Algarotti, *Sir Isaac Newton's Theory* (1742), 1.

70. Duperron de Castera trans., *Le Newtonianisme*, 1–2.

71. Agorni, "The Voice," 192.

72. In English-language editions, the "E" is followed by a long solid dash; in the Italian original, it is followed by three discrete dashes.

73. Algarotti, *Sir Isaac Newton's Theory* (1742), 2.

74. Duperron de Castera trans., *Le Newtonianisme*, 2.

75. As Gevirtz has noted, "Haywood translated into English Louis Adrien Duperron

de Castera's novel, *La Pierre Philosophale des Dames, ou les Caprices de l'Amour et du Destin, Nouvelle Historique* (1723), a work that associates alchemy and irrationality" (*Women, The Novel*, 103–4). For more on Haywood and reception, see also Orr, "Basis for Attribution," and Spedding, "Eliza Haywood's Readers."

76. Carter trans., *Newtonianism* (1739), 2.

77. Agorni, "The Voice," 192.

78. Carter trans., *Newtonianism* (1739), 4.

79. Algarotti, *Sir Isaac Newton's Theory* (1742), 15.

80. Carter trans., *Newtonianism* (1739), 11.

81. Algarotti, *Sir Isaac Newton's Theory* (1742), 13.

82. Fara notes an occurrence of Newton in a sexual fantasy from Denis Diderot's *Les Bijoux Indiscrètes* (1748), so this is not a unique midcentury instance of Newton's affiliation with sexualized representations (Fara, *Making of Genius*, 127). The sexualized changes to the marchioness by a French translator may also be a slight at her French inspiration, Du Châtelet.

83. Carter trans., *Newtonianism* (1739), 32–33. This passage continues, adopting the extended metaphor of sculpture and marble: philosophers like Newton "make observations and prepare materials for posterity to build systems upon, more fortunate in their duration, than those that preceded the present age" (33).

84. Carter trans., *Newtonianism* (1739), 33–34.

85. Pope, "Epitaph," 242.

86. Algarotti, *Sir Isaac Newton's Theory* (1742), 33.

87. *The Mind Is a Metaphor.* http://staging.metaphors.lib.virginia.edu/metaphors?q=marble&sort=relevance.

88. Ibid.

89. Woolfe, *Marbled Paper.* One other material connection between marble and print is, of course, the presence of paper marbling, which was a popular treatment of endpapers throughout the eighteenth century, and would have been present in at least some of the bindings of *Newtonianism for Ladies*. "Carving marble" would then be linked with bookbinding rather than publication.

90. Algarotti, *Sir Isaac Newton's Theory* (1742), 34.

91. Descartes proposed a comprehensive philosophical system in his *Principia Philosophiae;* however, *Il Newtonianismo per le dame* does not appear to regard Descartes's system as a system.

92. Lubey, *Excitable Imaginations,* 71.

93. Perry, *Novel Relations,* 237.

94. Carter trans., *Newtonianism* (1739), 9. Algarotti, *Il Newtonianismo per le dame* (1737), 5.

95. Algarotti, *Sir Isaac Newton's Theory* (1742), 10.

96. Earlier I cited Moretti's analysis of titles in "Style, Inc.," from *Distant Reading,* and his close attention to the rhetoric of book titles helps inform my analysis here.

97. Fellow of the Royal Society, Fellow of the Society of Antiquaries.

98. Moretti, "Style, Inc.," *Distant Reading*. Moretti shows how a title could be used to market or sell a work, including the role of a title's brevity. The 1765 edition offers a briefer title to appeal to a larger market. Condensing the title also expanded the genders to whom it could appeal.

99. Carter was not listed on the title page of the 1739 edition, but Cave publicized her translation widely.

100. Agorni, "The Voice," 192.

101. Sher, *Enlightenment and the Book*, 269.

102. Valenza explains that this process of translation had been taking place throughout the century, inspired by the publication of the *Principia*. However, I see the defined markets she argues are present at the *Principia*'s time of publication emerging later in the century rather than earlier, and never existing in isolation. Even with these later emerging specialized markets, one must not over-attribute social developments to the presence of printed popularizations, as my next chapter on the New York Society Library reveals. Readers at the NYSL read not only scientific texts but also popularizations and reference works that contained scientific information. Many readers borrowed all three kinds of book. Although there may have been expert writers, it does not logically follow that readers were themselves lacking in expertise, or only interested in reading original research.

103. Newbery also published Charlotte Palmer's novel *Female Stability*, one of the most popular works read at the NYSL during the 1789–92 period documented by the library's charging ledger, discussed in the next chapter.

104. See also my analysis of attribution in the *Principia* and the *Opticks* in chapter 2, as well as Cohen's article on the title page of the *Opticks*, also discussed in that chapter.

105. Secord, "Newton in the Nursery," and Langman, "Audience."

106. Martin, *Young Gentleman and Lady's Philosophy*.

107. Lennox, *Lady's Museum*.

108. Langman, "Audience," 52.

109. Fissell and Cooter, "Exploring Natural Knowledge," 138.

110. My next chapter will show how these audiences intersected with one another.

111. Guile, "Newtonian Pictures," 209.

112. Paula Findlen, Lead Researcher, "Francesco Algarotti," *Mapping the Republic of Letters*, Stanford University, http://republicofletters.stanford.edu/casestudies/algarotti.html.

113. Guile, "Newtonian Pictures," 209.

114. This title was also printed for Owen and Hope.

115. Algarotti, *Painting*, 5.

116. Algarotti, *Opera*, 21.

117. Carter trans., *Newtonianism* (1739), 210.

118. Ibid., 211.

119. Ibid., 211.

120. Ibid., 212.

121. Carter herself was painted as Minerva by John Fayram, c. 1735–41.

122. Carter trans., *Newtonianism* (1739), 213.

123. Patricia Howard, qtd. in *Italian Literature before 1900,* 867.

124. Algarotti, *Opera,* 1–2.

125. Ibid., 1.

126. Carter trans., *Newtonianism* (1739), viii.

127. Ibid., ix.

128. Ibid., ix.

129. Ibid., vii.

130. Algarotti, *Opera,* 2.

131. Ibid., 2.

132. Golinski analyzes the local foundations of experimental popularization, which parallels descriptions of theatrical popularity. As "inequalities of assets between scientists and lay people are not an unfortunate product of current institutional arrangements, but a precondition for the construction of natural knowledge," so too must opera have a community of experts and an array of resources to mount a production (*Science as Public Culture,* 3).

133. Algarotti, *Opera,* 4.

134. Carter trans., *Newtonianism* (1739), 26.

135. Ibid., xiv.

136. Algarotti, *Painting,* 17.

137. See Messbarger, *Lady Anatomist,* for more on Manzolini's work in eighteenth-century anatomy.

138. Carter trans., *Newtonianism* (1739), 5.

139. This volume was borrowed by one patron of the NYSL, James Inglis, who changed to borrowing the works of Littleton immediately afterward. He also borrowed Priestley's works, so he would not have been opposed to scientific content. He also borrowed volume 2 of the *Britannica* and a work on algebra immediately after returning it, so he may have been one of the readers of the *Britannica* who enjoyed its scientific content.

140. Anonymous, "Foreign Article," *Critical Review,* 230.

141. Ibid., 227.

142. Ibid., 227.

143. Fontenelle died at age 99 in 1757.

5. Reading Popular Newtonianism at the New York Society Library, 1789–1792

1. Parrish, *American Curiosity.*

2. Delbourgo, *Most Amazing Scene of Wonders.*

3. *City Readers, New York Society Library,* http://cityreaders.nysoclib.org/.

4. Glynn, *Reading Publics.*

5. I am indebted to other library history scholars for modeling ways to study readership by uniting online sources with other historical records. David Allan's work on eighteenth-century British readers and the various places they read socially—including book clubs and libraries—has helped me to understand the context of how books circulated, and what kinds of institutions facilitated that circulation. David Allan, *Making British Culture: English Readers and the Scottish Enlightenment, 1740–1830* (New York and Abingdon, OX: Routledge, 2008); *A Nation of Readers: The Lending Library in Georgian England* (London: British Library, 2008). Mark R. M. Towsey's *Reading the Scottish Enlightenment: Books and Their Readers in Provincial Scotland, 1750–1820* (Leiden: Brill, 2010) is another thorough study that helps to contextualize the kinds of books found in lending libraries across the eighteenth century.

6. Raven, *London Booksellers.*

7. Felsenstein and Connolly, *What Middletown Read.* The authors use library borrowing data in conjunction with other sources to reconstruct how the Muncie (Indiana) Public Library came together and how print consumption revealed itself in its nineteenth-century borrowing records.

8. Spedding, "Eliza Haywood's Eighteenth-Century Readers." This article also mentions my discussion of the library in an earlier essay: Miller, "Publishers and Gendered Readership." Furlong, "Libraries, Booksellers, and Readers."

9. Allan, *Nation of Readers,* 65. Allan accounts for some of the larger libraries by claiming libraries' ambitions and citizens' "civic pride" worked together to expand in size.

10. McClellan's *Science Reorganized* describes scientific and learned societies in eighteenth-century America, including the American Philosophical Society, the New York Society for Promoting Useful Knowledge, and the Boston Academy. In the 1970s, Lundberg and May performed an extensive quantitative study on American libraries in terms of popular holdings. Their work with the NYSL included sorting through its extant library catalogues and tabulating the numbers of books, but not the borrower data or volume-borrowing information. Their divisions of the Enlightenment and its books into several separate categories, including a "first, second, and third Enlightenment," and their exclusion of many important works also means that the study, while useful, could benefit from more comprehensive application (Lundberg and May, "Enlightened Reader in America").

11. "John Stevens," http://cityreaders.nysoclib.org/Detail/entities/1128.

12. It is important to acknowledge that different books were often borrowed on one account to circulate to different family members; Stevens, who died shortly after the ledger ends, was well past child-rearing years, and so it is likely that these books were borrowed for his personal use rather than to educate children at home.

13. "*Encyclopaedia Britannica,*" http://cityreaders.nysoclib.org/Detail/objects/3275.

14. "Gulian Verplanck," http://cityreaders.nysoclib.org/Detail/entities/1261.

15. Ibid.

16. "Malachi Treat," http://cityreaders.nysoclib.org/Detail/entities/1199.

17. Bowd, "Useful Knowledge or Polite Learning?"

18. "Garrett Abeel," http://cityreaders.nysoclib.org/Detail/entities/5.

19. "*Philosophical Transactions,*" http://cityreaders.nysoclib.org/Detail/objects/2347. An out-of-place 1726 volume appears to be mis-entered into the ledger here.

20. "Garrett Abeel," http://cityreaders.nysoclib.org/Detail/entities/5.

21. Aspray and Hayes, *Everyday Information.*

22. We should remember that many readers encountered Newton and his ideas not through explicitly scientific works, but through the numerous editions of *The Seasons* that reprinted Thomson's poem about Newton.

23. "Works of Buffon," http://cityreaders.nysoclib.org/Detail/objects/1297.

24. "John Kemp," http://cityreaders.nysoclib.org/Detail/entities/598; "William Samuel Johnson," http://cityreaders.nysoclib.org/Detail/entities/573.

25. Sher, *Enlightenment and the Book,* 449.

26. Ibid., 496.

27. Volume 2 was borrowed 62 times, vol. 3 was borrowed 21 times, and vol. 4 was borrowed 17 times.

28. Other works with scientific content were popular at the NYSL, including Lavater's *Physiognomie* (1772)—it was borrowed by 70 unique readers. Vols. 1 and 2 were checked out 34 times each and vol. 3 was borrowed 27 times ("*Physiognomie,*" http://cityreaders.nysoclib.org/Detail/objects/2179).

29. Fara, *Making of Genius,* 25.

30. *Encyclopaedia Britannica,* http://www.britannica.com.

31. Doig et al., "James Tytler's Edition."

32. Ibid., 155.

33. Sher, *Enlightenment and the Book,* 135.

34. Ibid., 135.

35. Yeo, *Encyclopedic Visions,* 70.

36. "James Watson, Jr.," http://cityreaders.nysoclib.org/Detail/entities/1293.

37. "Daniel Crommelin Verplanck," http://cityreaders.nysoclib.org/Detail/entities/1258.

38. "Gulian Verplanck," http://cityreaders.nysoclib.org/Detail/entities/1261. This was the volume that contained the entry on astronomy.

39. "Samuel Jones," http://cityreaders.nysoclib.org/Detail/entities/583; "John Cozine," http://cityreaders.nysoclib.org/Detail/entities/256.

40. "Valentine Peters," http://cityreaders.nysoclib.org/Detail/entities/894.

41. "This Day Is Published by Elliot & Kay."

42. Ibid., 8.

43. Ibid.

44. Ibid. This implies that the third edition was able to be typeset with relatively few materials, as only ten formes of type would have been necessary for each installment or "number."

45. Ibid.

46. Sher, *Enlightenment and the Book*, 556.

47. Ibid., 557.

48. Ibid., 558.

49. Ibid., 559.

50. Ibid.

51. Ibid., 561.

52. The copy I consulted is in the *Encyclopaedia Britannica* company archives in Chicago. Doig et al.'s essay is very thorough for those who seek additional information.

53. Newton himself was a subscriber to encyclopedias, including the *Lexicon Technicum*, to which he was also a contributor (Yeo, *Encyclopedic Visions*, 51). Yeo describes Newton's upward mobility on the subscriber list. By 1710 he was listed as "Sir Isaac Newton, Knight," "but still below other names, such as 'Colonel Nicholson, late Governor of Virginia'" (49). Newton also bought twelve copies of Henry Pemberton's *A View of Sir Isaac Newton's Natural Philosophy* (1728), though he did not live to see its publication.

54. Yeo, *Encyclopedic Visions*, 147.

55. Tytler et al., *Encyclopaedia Britannica*, 2nd ed., 1:224–37.

56. Ibid., 2:747–850.

57. Ibid., 2:901.

58. Ibid., 2:902.

59. "*Encyclopaedia Britannica*," http://cityreaders.nysoclib.org/Detail/objects/3275.

60. Glynn, "New York Society Library," 493. Glynn writes of science in the NYSL but does not discuss popularizations: "Science in 1758 was not yet the exclusive province of the professional scientist, and an enlightened gentleman was expected to be conversant in the scientific advances of an enlightened age" (502).

61. See Furlong for more information about the library's early acquisitions. I am also grateful to Rob Koehler for informally sharing his knowledge of the NYSL's practices.

62. Towsey, *Reading the Scottish Enlightenment*, 140.

63. Sher, *Enlightenment and the Book*, 496, 497.

64. "*Encyclopaedia Britannica*," http://cityreaders.nysoclib.org/Detail/objects/3275.

65. "Malachi Treat," http://cityreaders.nysoclib.org/Detail/entities/1199.

66. "Garrett Abeel," http://cityreaders.nysoclib.org/Detail/entities/5.

67. "*Encyclopaedia Britannica*," http://cityreaders.nysoclib.org/Detail/objects/3275.

68. Ibid.

69. Ibid.

70. Doig et al., "James Tytler's Edition," 101.

71. Yeo, *Encyclopedic Visions*, 162.

72. Doig et al., "James Tytler's Edition," 80.

73. Tytler et al., *Encyclopaedia Britannica*, 2nd ed., 2:1335–40. In comparison to Boyle's rather extensive biography as a scientific figure, literary figures receive shorter treatment. Aphra Behn, often neglected in discussions of the latter parts of the eighteenth century, nonetheless has a single-page entry (2:1088). Some noteworthy elements of this biography include the classification of *Oroonoko* as a novel and the description of Behn's writing career as a libertine one, in which she "devoted her life entirely to pleasure and the muses." The entry describes Behn as having written works of uneven quality, but does not discredit her because of her gender but rather because she wrote for money: "The best and perhaps the only excuse that can be made for [her works' inconsistent quality] is that, as she wrote for a livelihood, she was obliged to comply with the corrupt taste of the times" (2:1088). Samuel Richardson, also profiled, is described succinctly as a "celebrated English sentimental novel-writer" (9:6713). Richardson, however, comes across as somewhat less educated than Behn (who translated Fontenelle); his entry immediately lowballs him because he was "said to have understood no language but his own" (9:6713). In spite of these linguistic limitations, Richardson "acquired great reputation by his three epistolary novels, intitled *Pamela*, *Clarissa*, and *Sir Charles Grandison*; which show an uncommon knowledge of human nature" (9:6713). Daniel Defoe is described as "a writer famous for politics and poetry, [who] was bred a hosier" (4:2401). Defoe's *Robinson Crusoe* is referred to as plagiarism of one of its sources, the life of Selkirk. The origins of *Robinson Crusoe* are as follows, according to the encyclopedia: "[Selkirk] put the papers into the hands of De Foe, to digest for publication; who ungenerously converted the materials into the History of Robinson Crusoe, and returned Selkirk his papers again! A fraud for which, in a humane view, the distinguished merit of that romance can never atone. Daniel de Foe died at Islington, in 1731" (4:2401). This Defoe did not write either *A Journal of the Plague Year* (whose attribution status has not been confirmed) or the two novels whose attribution to Defoe is most frequently questioned: *Moll Flanders* and *Roxana*. Although the encyclopedia offered a concise biography of Behn and other writers like Richardson and Defoe, its coverage of the life of scientist Robert Boyle was far more extensive. The biographies of fiction writers are unsurprisingly short; in fact, it is surprising to see some of these writers listed at all, especially Behn. The *Encyclopaedia Britannica* entry implies that *Crusoe* was written by Defoe, who is best-known for poetry and politics, in contrast to the later Richardson, whose work with the early novel is documented.

74. Ibid., 2:1339.

75. Ibid., 2:1340.

76. Ibid., 4:2850. The edition of Euclid owned by the Society Library was actually Simson's Euclid, another Scottish-connected book.

77. Ibid., 5:3710–11.

78. Ibid., 5:3710.

79. Ibid., 7:5385.

80. Ibid., 7:5385.

81. Ibid., 7:5385.

82. Ibid., 7:5385.

83. Ibid., 7:5386.

84. See chapter 3 for more examples of these metaphors.

85. Ibid., 7:5385.

86. Ibid., 7:5386.

87. Ibid., 7:5385–86.

88. Ibid., 7:5387. Iliffe's forthcoming book on Newton's theology offers comprehensive information on Newton's documented theological stances as contrasted to the Anglican church.

89. Ibid., 7:5386–87.

90. Ibid., 7:5387.

91. Thomas Birch, Biography of Newton, from *A General Dictionary,* vol. 7 (1738). This biography has been reprinted and contextualized historically in Hall's *Sir Isaac Newton: Eighteenth-Century Perspectives*, 75–95.

92. Tytler et al., *Encyclopaedia Britannica,* 2nd ed., 7:5386.

93. Ibid., 7:5387.

94. Ibid., 7:5388.

95. Ibid., 7:5391.

96. Ibid., 7:5392.

97. Ibid., 7:5385–99.

98. Ibid., 7:5386.

99. Ibid., 7:5481.

100. Ibid., 7:5482.

101. Ibid., 7:5486.

102. Ibid., 7:5513.

103. New York encountered Newton in unexpected ways in this encyclopedia, not just in his biography. Although New York as a location was actually not listed under "N" in vol. 7, it could indeed be found in "Y" under "York" in vol. 10. This contradicts *The Great EB*'s assertion that New York was not actually present in the encyclopedia. The encyclopedia listings in "N" go directly from Newton—our Newton—to Niagara, bypassing New York.

104. Tytler et al., *Encyclopaedia Britannica,* 2nd ed., 1:236.

105. Ibid., 1:236.

106. Ibid., 3:1572.

107. Ibid., 4:3050–55.

108. Ibid., 2:747.

109. Ibid., 2:749.

110. Doig et al., "James Tytler's Edition," 104.

111. Ibid., 105.

112. Tytler et al., *Encyclopaedia Britannica,* 2nd ed., 2:766–67.

113. Doig et al., "James Tytler's Edition," 105.

114. Tytler et al., *Encyclopaedia Britannica,* 2nd ed., 2:771.

115. Ibid., 2:771.

116. Ibid., 2:778.

117. Doig et al., "James Tytler's edition," 105.

118. Ibid., 103.

119. Tytler et al., *Encyclopaedia Britannica,* 2nd ed., Marginal Notes 105 and 106, 2:781.

120. Ibid., 2:782.

121. Review of the second edition, qtd. in Doig et al., "James Tytler's Edition," 107.

122. Ibid., 108.

123. Tytler et al., *Encyclopaedia Britannica,* 2nd ed., 7:5003–40. This section had been controversial in the first edition, but most of it was reproduced for the second.

124. Doig et al., "James Tytler's Edition," 149.

125. Ibid., 148.

126. "Effingham Embree," http://cityreaders.nysoclib.org/Detail/entities/351.

127. Embree exemplifies the library readers Allan describes as "the thrusting and ambitious middle classes" who would be heavy borrowers and users of subscription libraries (*Nation of Readers,* 71).

128. "Effingham Embree," http://cityreaders.nysoclib.org/Detail/entities/351.

129. "John Kemp," http://cityreaders.nysoclib.org/Detail/entities/598.

130. Keep, *History of the New York Society Library,* 236.

131. McCaughey, *Stand, Columbia,* 59.

132. "John Kemp," http://cityreaders.nysoclib.org/Detail/entities/598. *Charging Ledger, 1793–1816,* Columbia University Rare Books and Manuscripts, call no. CPH.

133. "John Kemp," http://cityreaders.nysoclib.org/Detail/entities/598.

134. Raven, *London Booksellers,* 175.

135. Founded 1770 (ibid., 182).

136. Towsey mentions that Maclaurin's *Account of Newton* as well as Ferguson's *Astronomy* were very popular in Scottish subscription libraries, so the Scottish Kemp would have had access to the same works as his Scottish contemporaries (Towsey, *Reading the Scottish Enlightenment,* 67).

137. Robins was a protégé of Henry Pemberton; this book includes research on both "gunnery, and its sister art, fortification" (*Mathematical Tracts* [1761]), 5).

138. Newton's connections helped ultimately link Maclaurin with his job in Edinburgh, where Newton is said to have offered to pay Maclaurin's salary himself. Grabiner describes Maclaurin's use of what Cohen called "the Newtonian style" of mathematics ("Newton, Maclaurin").

139. Although this has been a book largely about Newton, there were many mathematical changes following Euler that Kemp needed to be aware of. Guicciar-

dini's assessment of the changes to mathematics in midcentury provides an excellent summary:

> From 1730 to 1750 the concept of function emerged, and multivariate functions, partial differential equations, and the calculus of variations were first defined and studied. Mechanics too, as Truesdell has convincingly maintained, underwent a profound transformation into a science, not envisaged by Newton, based on minimum principles and applied to extended (rigid, flexible, and fluid) bodies. The birth of what can be called the 'Eulerian calculus and mechanics,' a process which was due mainly, but not exclusively, to the Continentals, marks a deep transformation of the mathematical sciences. After this change, the *Principia* and the debate surrounding it belong to the past. (Guicciardini, *Reading the "Principia,"* 12)

140. *Charging Ledger, 1793–1816*, Columbia University Rare Books and Manuscripts, call no. CPH.

141. McCaughey, *Stand, Columbia*, 59.

142. "Peter Wilson," http://cityreaders.nysoclib.org/Detail/entities/1334.

143. "William Samuel Johnson," http://cityreaders.nysoclib.org/Detail/entities/573.

144. Williams, *Significance of the Printed Word*. 10.

145. Glynn, "New York Society Library," 509.

146. Tytler et al., preface to *Encyclopaedia Britannica*, 2nd ed.

147. Gantt and Williams, "Psychology and the Legacy of Newtonianism," 83–100, 95.

148. "Jan. 4, 2010," *Google Doodle*, https://www.google.com/doodles/sir-isaac-newtons-367th-birthday.

Bibliography

"Advertisement." *Philosophical Transactions of the Royal Society* 16 (1686–92): 297.

Agamben, Giorgio. *Potentialities: Collected Essays in Philosophy*. Stanford, CA: Stanford University Press, 2000.

Agorni, Mirella. "The Voice of the 'Translatress': From Aphra Behn to Elizabeth Carter." *Yearbook of English Studies* 28 (1998): 181–95.

Aikin, John. *An Essay on the Application of Natural History to Poetry*. Warrington, UK: Printed by W. Eyres for J. Johnson, 1777.

Albury, W. R. "Halley's Ode on the *Principia* of Newton and the Epicurean Revival in England." *Journal of the History of Ideas* 39, no. 1 (1978): 24–43.

Algarotti, Francesco. *An Essay on Painting Written in Italian*. London: Davis and Reymers, 1762. Originally published as *Saggio sopra la pittura*.

———. *An Essay on Opera Written in Italian by Count Algarotti*. London: Davis and Reymers, 1755. Originally published as *Saggio sopra l'opera in musica*.

———. *Il Newtonianismo per le dame*. Naples, n.p.: 1737.

———. *Le Newtonianisme pour les dames: ou, Entretiens sur la lumière, sur les couleurs, et sur l'attraction*. Translated by Louis-Adrien Duperron de Castera. Paris: Montalant, 1738.

———. *Sir Isaac Newton's Philosophy Explain'd for the Use of the Ladies in Six Dialogues on Light and Colours*. Translated by Elizabeth Carter. London: Edward Cave, 1739.

———. *Sir Isaac Newton's Theory of Light and Colours, and His Principle of Attraction, Made Familiar to the Ladies*. London: George Hawkins, 1742.

———. *The Lady's Philosophy: or Sir Isaac Newton's Theory of Light and . . . Made Familiar to the Ladies in Several Entertainments*. London: Francis Newbery, 1772.

———. *The Philosophy of Sir Isaac Newton Explained, in Six Dialogues on Light and Colors*. Glasgow: Robert Urie, 1765.

Allan, David. *Making British Culture: English Readers and the Scottish Enlightenment, 1740–1830*. New York: Routledge, 2008.

———. *A Nation of Readers: The Lending Library in Georgian England*. London: British Library, 2008.

Andersen, Jennifer, and Elizabeth Sauer, eds. *Books and Readers in Early Modern England: Material Studies*. Philadelphia: University of Pennsylvania Press, 2002.

Anderson, Benedict. *Imagined Communities: Reflections on the Origins and Spread of Nationalism*. New York: Verso Books, 1983.

Arato, Franco. "Intorno al *Newtonianismo.* Quattro lettere inedite di Francesco Algarotti." *Giornale storico della letteratura italiana* 164, no. 528 (1987): 557–69.

———."Minerva and Venus: Algarotti's Newton's Philosophy for the Ladies." In *Men, Women, and the Birthing of Modern Science*, edited by Judith P. Zinsser, 111–22. Dekalb: Northern Illinois University Press, 2005.

Armintor, Deborah Needleman. "The Sexual Politics of Microscopy in Brobdingnag." *SEL: Studies in English Literature, 1500–1900* 47, no. 3 (2007): 619–40.

Aspray, William, and Barbara M. Hayes, eds. *Everyday Information: The Evolution of Information Seeking in America*. Cambridge, MA: MIT Press, 2011.

Bannet, Eve Tavor. *Transatlantic Stories and the History of Reading, 1720–1810: Migrant Fictions.* Cambridge: Cambridge University Press, 2011.

Barchas, Janine. *Graphic Design and the Eighteenth-Century Novel.* Cambridge: Cambridge University Press, 2008.

Barker, Francis. *The Tremulous Private Body: Essays on Subjection*. Ann Arbor: University of Michigan Press, 1984.

Barnes, A. W. *Post-Closet Masculinities in Early Modern England*. Lewisville, PA: Bucknell University Press, 2009.

Bender, John. "Enlightenment Fiction and the Scientific Hypothesis." In *Ends of Enlightenment*, 38–56. Stanford, CA: Stanford University Press, 2012.

———, and Michael Marrinan. *The Culture of Diagram*. Palo Alto, CA: Stanford University Press, 2010.

Berry, Helen. *Gender, Culture, and Society in Late Stuart England: The Cultural World of the "Athenian Mercury."* Burlington, VT: Ashgate, 2003.

Biagioli, Mario. *Galileo, Courtier: The Practice of Science in the Culture of Absolutism*. Chicago: University of Chicago Press, 1993.

———, and Peter Galison, eds. *Scientific Authorship*. Abingdon, UK: Taylor and Francis, 2003.

Birch, Thomas. "Isaac Newton." In *A General Dictionary, Historical and Critical, edited by* Pierre Bayle, 7:776–802. London: James Bettenham, 1738.

Blair, Ann. *Too Much to Know: Managing Scholarly Information before the Modern Age*. New Haven, CT: Yale University Press, 2010.

Bowd, Rebecca. "Useful Knowledge or Polite Learning? A Reappraisal of Approaches to Subscription Library History." *Library and Information History* 29, no. 3 (2013): 182–95.

Boyle, Frank. *Swift as Nemesis: Modernity and its Satirist*. Palo Alto, CA: Stanford University Press, 2000.

Brewer, David. *The Afterlife of Character, 1726–1825*. Philadelphia: University of Pennsylvania Press, 2005.

Buchwald, Jed Z., and Mordechai Feingold. *Newton and the Origin of Civilization*. Princeton, NJ: Princeton University Press, 2012.

Butler, Samuel. "The Elephant in the Moon." In *The Poetical Works of Samuel Butler*, 3:9–27. Edinburgh: Apollo Press, 1784.

Carter, Philip. *Men and the Emergence of Polite Society: Britain 1660–1800*. New York: Longman/St. Martin's Press, 1997.

Chartier, Roger. *Forms and Meanings: Texts, Performances, and Audiences from Codex to Computer*. Philadelphia: University of Pennsylvania Press, 1995.

———. "Laborers and Voyagers: From the Text to the Reader." *diacritics* 22, no. 2 (1992): 49–61.

Chico, Tita. "Gimcrack's Legacy: Sex, Wealth, and the Theater of Experimental Philosophy." *Comparative Drama* 41, no. 1 (2009): 29–49.

Cohen, I. Bernard. "The Case of the Missing Author: The Title Page of Newton's *Opticks* (1704) with Notes on the Title Page of Huygens's *Traité de la Lumière*." In *Isaac Newton's Natural Philosophy*, edited by Jed Z. Buchwald and I. Bernard Cohen, 15–46. Dibner Institute Series in the History of Science and Technology. Cambridge, MA: MIT Press, 2001.

———. "A Guide to Newton's *Principia*." In *Philosophiae Naturalis Principia Mathematica* [*The Mathematical Principles of Natural Philosophy, 1687*], by Isaac Newton, translated by I. Bernard Cohen and Anne Whitman. 1999. Reprint, Berkeley: University of California Press, 2016.

———. *Introduction to Newton's "Principia."* Cambridge, MA: Harvard University Press, 1971; Berkeley: University of California Press, 1999 and 2016.

Cohen, Michèle. "French Conversation or 'Glittering Gibberish'? Learning French in Eighteenth-Century England." In *Didactic Literature in England: 1500–1800*, edited by Natasha Glaisyer and Sara Pennell, 99–117. Hampshire, UK: Ashgate, 2003.

Coppola, Al. "Retraining the Virtuoso's Gaze: Behn's *Emperor of the Moon*, the Royal Society, and the Spectacles of Science and Politics." *Eighteenth-Century Studies* 41, no. 4 (2008): 481–506.

———*The Theater of Experiment: Staging Natural Philosophy in Eighteenth-Century Britain*. New York: Oxford University Press, 2016.

Costa, Shelley. "The *Ladies' Diary:* Gender, Mathematics, and Civil Society in Early Eighteenth-Century England." *Osiris* 17 (2002): 49–73.

Cowley, Abraham. "To the Royal Society." *The Abraham Cowley Text and Image Archive*. http://cowley.lib.virginia.edu/works/RoyalSociety.htm.

"Daniel Crommelin Verplanck." City Readers, *New York Society Library*. http://cityreaders.nysoclib.org/Detail/entities/1258.

Darnton, Robert. "What Is the History of Books?" *Daedalus* 111, no. 3 (1982): 65–83.

Dear, Peter. *Discipline and Experience*. Chicago: University of Chicago Press, 1995.

———."Totius in Verba: Rhetoric and Authority in the Early Royal Society." *Isis* 76 (1985): 144–61.

Defoe, Daniel. *A Journal of the Plague Year*. London: Nutt, Roberts, Dodd, and Graves, 1722.

Davy de la Fautrière, Louis. "Épître Newtonienne." 1739. http://gallica.bnf.fr/ark:/12148/bpt6k5434537s.

Delbourgo, James. *A Most Amazing Scene of Wonders: Electricity and Enlightenment in Early America*. Cambridge, MA: Harvard University Press, 2006.

Derrida, Jacques. "Typewriter Ribbon: Limited Ink." In *Without Alibi*, translated by Peggy Kamuf, 71–160. Stanford, CA: Stanford University Press, 2002.

Desaguliers, John Theophilus. *The Newtonian System of the World, The Best Model of Government: An Allegorical Poem*. Westminster/London: Campbell and Roberts, 1728.

Descartes, René. *Principia Philosophiae*. Amsterdam: Louis Elzevir, 1644.

———. *Principia Philosophiae*. Amsterdam: Ex Typographia Blaviana, 1692.

Dobbs, Betty Jo Teeter. *The Foundations of Newton's Alchemy, or, the Hunting of the Greene Lyon*. Cambridge: Cambridge University Press, 1983.

Doig, K. Hardesty, F. A. Kafker, J. Loveland, and D. A. Trinkle. "James Tytler's Edition: A Vast Expansion and Improvement." In *The Early Britannica: The Growth of an Outstanding Encyclopedia*, edited by F. A. Kafker and J. Loveland, 69–155. *Studies in Voltaire and the Eighteenth Century*. Oxford: Voltaire Foundation, 2009.

Dry, Sarah. *The Newton Papers*. Oxford: Oxford University Press, 2014.

"Effingham Embree." *City Readers, New York Society Library*. http://cityreaders.nysoclib.org/Detail/entities/351.

Eisenstein, Elizabeth. *The Printing Press as an Agent of Change*. Cambridge: Cambridge University Press, 1979.

"*Encyclopaedia Britannica*." *City Readers*, New York Society Library. http://cityreaders.nysoclib.org/Detail/objects/3275.

Fara, Patricia. *Newton: The Making of Genius*. New York: Columbia University Press, 2004.

Farooq, Jennifer. *Preaching in Eighteenth-Century London*. Woodbridge, UK: Boydell and Brewer, 2013.

Feingold, Mordechai. "Mathematicians and Naturalists." In *Isaac Newton's Natural Philosophy*, edited by Jed Z. Buchwald and I. Bernard Cohen, 77–102. Boston: MIT Press, 2004.

———. *The Newtonian Moment: Isaac Newton and the Making of Modern Culture*. Oxford: Oxford University Press, 2004.

Felsenstein, Frank, and James J. Connolly. *What Middletown Read: Print Culture in an American Small City*. Amherst: University of Massachusetts Press, 2015.

Findlen, Paula. "Translating the New Science: Women and the Circulation of Science in Enlightenment Italy." *Configurations* 3, no. 2 (1995): 167–206.

Fissell, Mary, and Roger Cooter. "Exploring Natural Knowledge: Science and the Popular." In *Eighteenth-Century Science*, vol. 4 of *The Cambridge History of Science*, edited by Roy Porter, 129–58. Cambridge: Cambridge University Press, 2003.

Flamsteed, John. Suppressed preface to the *Historia Coelestis Britannica*. *The Newton Project*. http://www.newtonproject.ox.ac.uk/view/texts/normalized/OTHE00033.

Fontenelle, Bernard Le Bovier de. *Conversations on the Plurality of Worlds.* 1687. Translated by Nina Rattner Gelbart. Berkeley: University of California Press, 1990.

———. *Eloge de M. Neuton.* 1728. http://gallica.bnf.fr/ark:/12148/bpt6k5400744j.

———. *Entretiens sur la pluralité des mondes.* Paris: Michel Guerout, 1687.

Force, James E., and Sarah Hutton, eds. *Newton and Newtonianism: New Studies.* Dordrecht: Kluwer, 2004.

Furlong, Jennifer. "Libraries, Booksellers, and Readers: Changing Tastes at the New York Society Library in the Long Eighteenth Century." *Library and Information History* 31, no. 3 (2015): 198–212.

Gabbard, D. Christopher. "Clashing Masculinities in Aphra Behn's *The Dutch Lover.*" *SEL: Studies in English Literature, 1500–1900* 4, no. 3 (2007): 557–72.

Gaine, Hugh. *The Charter, Bye-Laws, and Names of the Members of the New-York Society Library, with a Catalogue of the Books Belonging to the Said Library.* New York: Hugh Gaine, 1789.

Galilei, Galileo. *Dialogo sopra I due massimi sistemi del mondo tolemaico e copernicano.* Fiorenza: Gio. Batista Landini, 1632.

Gantt, Edwin E., and Richard N. Williams. "Psychology and the Legacy of Newtonianism: Motivation, Intentionality, and the Ontological Gap." *Journal of Theoretical and Philosophical Psychology* 34, no. 2 (2014): 83–100.

"Garrett Abeel." *City Readers, New York Society Library.* http://cityreaders.nysoclib.org/Detail/entities/5.

Gedde, John. *A New Discovery of an Excellent Method of Bee-Houses & Colonies.* London, 1675.

Genette, Gérard. *Paratexts: Thresholds of Interpretation.* Cambridge: Cambridge University Press, 1997.

Gevirtz, Karen. *Women, the Novel, and Natural Philosophy, 1660–1727.* Basingstoke, UK: Palgrave, 2014.

Gingerich, Owen. *The Book Nobody Read: Chasing the Revolutions of Nicolaus Copernicus.* New York: Walker, 2004.

Ginsberg, Allen. *Howl I, "Howl" and Other Poems.* San Francisco: City Lights, 1956.

Girten, Kristin M. "Mingling with Matter: Tactile Microscopy and the Philosophic Mind in Brobdignag and Beyond." *Eighteenth Century: Theory and Interpretation* 54, no. 4 (Winter 2013): 497–520.

Glynn, Tom. "The New York Society Library: Books, Authority, and Publics in Colonial and Early Republican New York." *Libraries and the Cultural Record* 40, no. 4 (2005): 493–529.

———. *Reading Publics: New York City's Public Libraries, 1754–1911.* New York: Fordham University Press, 2015.

Golinski, Jan. *Science as Public Culture: Chemistry and Enlightenment in Britain, 1760–1820.* Cambridge: Cambridge University Press, 1992, 1999.

Grabiner, Judith V. "Newton, Maclaurin, and the Authority of Mathematics." *Mathematical Association of America Monthly* (December 2004): 841–52.

Grafton, Anthony. *Bring Out Your Dead: The Past as Revelation*. Cambridge, MA: Harvard University Press, 2001.

———. *The Footnote: A Curious History*. Cambridge, MA: Harvard University Press, 1997.

Gray, George John. *A Bibliography of the Works of Sir Isaac Newton, Together with a List of Books Illustrating His Works*. Cambridge: Macmillan and Bowes, 1888.

Guerrini, Anita. "The 'Virtual Menagerie': The *Histoire des Animaux* Project." *Configurations* 14 (2006): 29–41.

Guicciardini, Niccolò. *Isaac Newton on Mathematical Certainty and Method*. Cambridge, MA: MIT Press, 2009.

———. *Reading the "Principia."* Cambridge: Cambridge University Press, 2003.

———. "The Role of Musical Analogies in Newton's Optical and Cosmological Work." *Journal of the History of Ideas* 74, no. 1 (2013): 45–67.

Guile, Carolyn. "Newtonian Pictures: Francesco Algarotti's Writings on Art as Science." In *Reflections on Renaissance Venice. Essays in Honor of Patricia Fortini Brown*, 209–20. Milan: 5 Continents, 2013.

"Gulian Verplanck." *City Readers, New York Society Library*. http://cityreaders.nysoclib.org/Detail/entities/1261.

Habermas, Jürgen. *The Structural Transformation of the Public Sphere: An Inquiry into a Category of Bourgeois Society*. Translated by Thomas Burger and Frederick Lawrence. Cambridge, MA: MIT Press, 1991.

Hall, Alfred Rupert. "Thomas Birch's Article." In *Sir Isaac Newton: Eighteenth-Century Perspective*, 75–95. Oxford: Oxford University Press, 1999.

Hall, Marie Boas. *Henry Oldenburg: Shaping the Royal Society*. Oxford: Oxford University Press, 2002.

Halley, Edmond. "Ode on This Splendid Ornament of Our Time and Our Nation, the Mathematico-Physical Treatise by the Eminent Isaac Newton." In *Philosophiae Naturalis Principia Mathematica* [*The Mathematical Principles of Natural Philosophy, 1687*], by Isaac Newton, translated by I. Bernard Cohen and Anne Whitman. Berkeley: University of California Press, 2016.

———. Review of the *Principia*, from "Accounts of Books." *Philosophical Transactions of the Royal Society* 16 (1686): 291–97.

Hammond, Paul. *Figuring Sex between Men from Shakespeare to Rochester*. Oxford: Oxford University Press, 2002.

Haraway, Donna. *Modest_Witness@Second_Millennium.FemaleMan_Meets OncoMouse: Feminism and Technoscience*. New York: Routledge, 1997.

Harding, Sarah G. *The Science Question in Feminism*. Ithaca, NY: Cornell University Press, 1986.

Harrison, John. *The Library of Isaac Newton*. Cambridge: Cambridge University Press, 1978, 2008.

Hartman, Anne. "Confessional Counterpublics in Frank O'Hara and Allen Ginsberg." *Journal of Modern Literature* 28, no. 4 (Summer 2005): 40–56.

Hawley, Judith. "Elizabeth Carter." *Oxford Dictionary of National Biography*. http://www.oxforddnb.com/index/4/101004782/.

Healey, Robin, ed. *Italian Literature before 1900 in English Translation: An Annotated Bibliography*. Toronto: University of Toronto Press, 2011.

Hooke, Robert. *Micrographia: or Some Physiological Descriptions of Minute Bodies Made by Magnifying Glasses, with Observations and Inquiries Thereupon*. London: Martyn and Allestry, 1665.

Horne, George, Bishop of Norwich. "A Satirical Letter, Written When a Student at Oxford, but Unsigned and Unaddressed, on the Scientific Views of Sir Isaac Newton, etc." *British Library Add. MS 34079: 1513–1839* ("Old Alfred's Walls" [University Coll., Oxford], Nov. 8, 1749).

Hunter, Michael. *Establishing the New Science: The Experience of the Early Royal Society*. Woodbridge, UK: Boydell Press, 1995.

Hutchison, Keith. "What Happened to Occult Qualities in the Scientific Revolution?" *Isis* 73, no. 2 (1982): 233–53.

Hutton, Sarah. "Before *Frankenstein*." In *The New Science and Women's Literary Discourse: Prefiguring Frankenstein*, edited by Judy A. Hayden, 17–28. New York: Palgrave, 2011.

———. "Women, Science, and Newtonianism: Emilie du Châtelet versus Francesco Algarotti." In *Newton and Newtonianism: New Studies*, edited by James E. Force and Sarah Hutton, 183–203. *International Archives of the History of Ideas, vol.* 188. New York: Kluwer, 2004.

Iliffe, Robert. "Butter for Parsnips: Authorship, Audience, and the Incomprehensibility of the Principia." In *Scientific Authorship*, edited by Mario Biagioli and Peter Galison, 33–66. New York: Routledge, 2003.

———. "'Is he like other men?' The Meaning of the *Principia Mathematica* and the Author as Idol." In *Culture and Society in the Stuart Restoration*, edited by Gerald Maclean, 159–76. Cambridge: Cambridge University Press, 1995.

Iser, Wolfgang. *The Act of Reading: A Theory of Aesthetic Response*. Baltimore: John Hopkins University Press, 1974.

Jacob, Margaret C., and B. J. T. Dobbs. *Newton and the Culture of Newtonianism*. Amherst, NY: Humanity Books, 1995.

Jacob, Margaret C., and Larry Stewart. *Practical Matter: Newton's Science in the Service of Industry and Empire, 1687–1851*. Cambridge, MA: Harvard University Press, 2006.

"James Watson, Jr." *City Readers, New York Society Library*. http://cityreaders.nysoclib.org/Detail/entities/1293.

"Jan 4, 2010." *Google Doodle*. https://www.google.com/doodles/sir-isaac-newtons-367th-birthday.

Jardine, Lisa. *The Curious Life of Robert Hooke: The Man Who Measured London*. New York: Harper Collins, 2003.

"John Kemp." *City Readers, New York Society Library*. http://cityreaders.nysoclib.org/Detail/entities/598.

"John Cozine." *City Readers, New York Society Library.* http://cityreaders.nysoclib.org/Detail/entities/256.

"John Stevens." *City Readers, New York Society Library.* http://cityreaders.nysoclib.org/Detail/entities/1128.

Johns, Adrian. *The Nature of the Book: Print and Knowledge in the Making.* Chicago: University of Chicago Press, 1998.

———. *Piracy: The Intellectual Property Wars from Gutenberg to Gates.* Chicago: University of Chicago Press, 2010.

Johnson, James William. *A Profane Wit: The Life of John Wilmot, Earl of Rochester.* Rochester, NY: University of Rochester Press, 2004.

Jonson, Ben. "To the Memory of My Beloved the Author, William Shakespeare." In front matter to *Mr. William Shakespeares Comedies, Histories & Tragedies.* London: Jaggard and Blount, 1623.

Jung, Sandro. *James Thomson's "The Seasons," Print Culture, and Visual Interpretation, 1730–1842.* Lewisville, PA: Bucknell University Press, 2015.

Keep, Austin Baxter. *History of the New York Society Library.* New York: De Vinne Press, 1908.

Keynes, Milo. *The Iconography of Sir Isaac Newton to 1800.* Woodbridge, UK: Boydell Press, 2005.

Klein, Lawrence E. "Gender and the Public/Private Distinction in the Eighteenth Century: Some Questions about Evidence and Analytic Procedure." *Eighteenth-Century Studies* 29, no. 1 (1995): 97–109.

Kramnick, Jonathan. *Actions and Objects from Hobbes to Richardson.* Stanford, CA: Stanford University Press, 2010.

Kuhn, Thomas. *The Structure of Scientific Revolutions.* Chicago: University of Chicago Press, 1962.

Langman, Pete. "The Audience Is Listening: Reading Writing about Learning by Doing." In *Learning by Doing: Experiments and Instruments in the History of Science Teaching,* edited by P. Heering and R. Wittje, 31–54. Stuttgart: Franz Steiner, Verlag, 2011.

Lanser, Susan. *The Sexuality of History: Modernity and the Sapphic, 1565–1830.* Chicago: University of Chicago Press, 2014.

Latour, Bruno. *Reassembling the Social: An Introduction to Actor-Network-Theory.* Oxford: Oxford University Press, 2007.

Lennox, Charlotte. *The Lady's Museum: By the Author of "The Female Quixote."* London: Printed for J. Newbery in St. Paul's Church-Yard and J. Coote in Paternoster-Row, 1760–61.

Loveland, Jeff. "French Thought in William Smellie's Natural History: A Scottish Reception of Buffon and Condillac." In *Scotland and France in the Enlightenment,* edited by Deidre Dawson, Pierre Morère, and Kathleen Hardesty Doig, 192–220. Lewisburg, PA: Bucknell University Press, 2004.

Lualdi, Katharine J. *Sources of "The Making of the West." Vol. 2, Since 1500: Peoples and Cultures.* Boston: Bedford/St. Martin's, 2012.

Lubey, Kathleen. *Excitable Imaginations: Eroticism and Reading in Britain, 1660–1760.* Lewisburg, PA: Bucknell University Press, 2012.

Lundberg, David, and Henry F. May. "The Enlightened Reader in America." *American Quarterly* 28, no. 2 (1976): 262–93.

"Malachi Treat." *City Readers, New York Society Library.* http://cityreaders.nysoclib.org/Detail/entities/1199.

Mannheimer, Katherine. "Celestial Bodies: Readerly Rapture as Theatrical Spectacle in Aphra Behn's *The Emperor of the Moon.*" *Restoration: Studies in English Literary Culture, 1660–1700* 35, no. 1 (2011): 39–60.

Markley, Robert. *Fallen Languages: Crises of Representation in Newtonian England, 1660–1740.* Ithaca, NY: Cornell University Press, 1993.

Martin, Benjamin. *The Young Gentleman and Lady's Philosophy: In a Continued Survey of the Works of Nature and Art by Way of Dialogue.* London: W. Owen and B. Martin, 1759–63.

Maurer, Shawn Lisa. *Proposing Men: Dialectics of Gender and Class in the Eighteenth-Century English Periodical.* Stanford: Stanford University Press, 1998.

Mazzotti, Massimo. "Newton for Ladies." Introduction to edited text of *Il Newtonianismo per le dame. Bologna Science Classics Online.* University of Bologna, Department of Philosophy. http://137.204.24.205/cis13b/bsc03/intro_opera.asp?id_opera=32.

McClellan, James E. *Science Reorganized: Scientific Societies in the Eighteenth Century.* New York: Columbia University Press, 1985.

McCaughey, Robert. *Stand, Columbia: A History of Columbia University in the City of New York, 1754–2004.* New York: Columbia University Press, 2003.

McGeary, Thomas. "Clarissa Harlowe's 'Ode to Wisdom': Composition, Publishing History, and the Semiotics of Printed Music." *Eighteenth-Century Fiction* 24, no. 3 (2012): 431–58.

McGill, Meredith. "Copyright and Intellectual Property: The State of the Discipline." *Book History* 16, no.1 (2013): 387–427.

Messbarger, Rebecca. *The Lady Anatomist: The Life and Work of Anna Morandi Manzolini.* Chicago: University of Chicago Press, 2010.

Miller, Laura. "Masculinity, Space, and Late Seventeenth-Century Alchemical Practices." In *Gender and Space in Britain, 1660–1820,* edited by Karen Gevirtz and Mona Narain, 165–78. Farnum, UK: Ashgate, 2014.

———. "Publishers and Gendered Readership in English-Language Editions of *Il Newtonianismo per le Dame.*" *Studies in Eighteenth-Century Culture* 42 (2013): 191–214.

Moretti, Franco. "Style, Inc. Reflections on Seven Thousand Titles." In *Distant Reading,* 179–210. London: Verso, 2013.

Newbery, John [Tom Telescope, pseud.]. *The Newtonian System of Philosophy, Adapted to the Capacity of Young Gentlemen and Ladies.* London: John Newbery, 1761.

Newman, William R. *Atoms and Alchemy: Chymistry and the Experimental Origins of the Scientific Revolution.* Chicago: University of Chicago Press, 2006.

———, and Lawrence M. Principe. *Alchemy Tried in the Fire: Starkey, Boyle, and the Fate of Helmontian Chymistry.* Chicago: University of Chicago Press, 2002.

Newton, Isaac. "An Account of the Systeme of the World Described in Mr. Newton's Mathematicall Principles of Philosophy." *MS Add. 4005,* ff. 39–42. Miscellaneous Papers, Cambridge University Library. https://cudl.lib.cam.ac.uk/view/MS-ADD-04005/131.

———. "Draft Chapters of a Treatise on the Origin of Religion and Its Corruption." *Yahuda Ms. 41, The Newton Project.* http://www.newtonproject.ox.ac.uk/view/texts/normalized/THEM00077.

———. *Commercium Epistolicum.* London: Tonson and Watts, 1712.

———. *The Correspondence of Isaac Newton.* Edited by H. W. Turnbull et al. 7 vols. Cambridge: Cambridge University Press, 1959–77.

———. "A Letter from Mr. Isaac Newton . . . containing his New Theory about Light and Colors." *Philosophical Transactions of the Royal Society* 80 (Feb. 19, 1671/72): 3075–87.

———. "Letter to Halley," Oct. 18, 1686. *The Newton Project.* http://www.newtonproject.sussex.ac.uk/view/texts/normalized/NATP00328.

———. Manuscript dated May 16, 1666. In *The Mathematical Papers of Isaac Newton, vol. 1, 1664–1666,* edited by D. T. Whiteside, 392–99. Cambridge: Cambridge University Press, 1981.

———. *"Paradoxical Questions Concerning the Morals and Actions of Athanasius and His Followers."* Manuscript N563M3 P222, William Andrews Clark Memorial Library, Los Angeles, CA.

———. *Philosophiae Naturalis Principia Mathematica.* London: Joseph Streater, 1687.

———. *Philosophiae Naturalis Principia Mathematica.* Cambridge: n.p., 1713.

———. *Philosophiae Naturalis Principia Mathematica.* London: William and John Innys, 1726.

———. *Philosophiae Naturalis Principia Mathematica.* Translated by Andrew Motte. London: Benjamin Motte, 1729.

———. *Philosophiae Naturalis Principia Mathematica.* Translated by I. Bernard Cohen and Anne Whitman. Berkeley: University of California Press, 2016.

———. "Seven Drafts of Newton's Defence of the Chronology of Ancient Kingdoms." *The Newton Project.* http://www.newtonproject.ox.ac.uk/view/texts/normalized/THEM00070.

New York Society Library. Charging Ledger, 1793–1816. Columbia University Rare Books and Manuscripts, call no. CPH.

Nicolson, Marjorie Hope. *Newton Demands the Muse: Newton's "Opticks" and the Eighteenth Century Poets.* Princeton, NJ: Princeton University Press, 1946.

"Notes on the editing of the rest of the manuscript." *The Newton Project*. http://www.newtonproject.ox.ac.uk/view/texts/normalized/THEM00408.

Orr, Leah. "The Basis for Attribution in the Canon of Eliza Haywood." *The Library* 12, no. 4 (2011): 335–75.

Ovid. *Metamorphoses* (Francofurdi: Ex officina typographica Ioannis Wecheli, 1593). Burndy Library Collection, Huntington Library, San Marino, CA.

———. *Ovid's "Metamorphosis" Englished, by Geo. Sandys*. Translated by George Sandys. London: E. F. for G. Sawbridge, A. Roper, T. Basset, J. Wright, and R. Chiswell, 1678.

Palmer, Charlotte. *Female Stability*. London: F. Newbery, 1780.

Parker, Blanford. *The Triumph of Augustan Poetics: English Literary Culture from Butler to Johnson*. Cambridge: Cambridge University Press, 2006.

Parrish, Susan Scott. *American Curiosity: Cultures of Natural History in the Colonial British Atlantic World*. Chapel Hill: University of North Carolina Press, 2006.

Pemberton, Henry. *View of Sir Isaac Newton's Philosophy*. London: S. Palmer, 1728.

Pepys, Samuel. "Diary Entries from February 1665." *The Diary of Samuel Pepys* (website). http://www.pepysdiary.com.

Perrault, Claude. *Memoires pour servir à l'histoire naturelle des animaux*. Translated by Alexander Pitfeild [*sic*]. London: Joseph Streater, 1688.

Perry, Ruth. *Novel Relations: The Transformation of Kinship in English Literature and Culture, 1748–1818*. Cambridge: Cambridge University Press, 2004.

"Peter Wilson." *City Readers, New York Society Library*. http://cityreaders.nysoclib.org/Detail/entities/1334.

"*Philosophical Transactions*." City Readers, *New York Society Library*. http://cityreaders.nysoclib.org/Detail/objects/2347.

"*Philosophical Transactions* in the Seventeenth Century." *Publishing the Philosophical Transactions* (website). https://arts.st-andrews.ac.uk/philosophicaltransactions/brief-history-of-phil-trans/phil-trans-in-1673/.

"*Physiognomie*." *City Readers, New York Society Library*. http://cityreaders.nysoclib.org/Detail/objects/2179.

Piper, Andrew. *Dreaming in Books*. Chicago: University of Chicago Press, 2009.

———."Footnote Detection." *txtlab@McGill (website)*, Sept. 22, 2014. http://txtlab.org/?p=395.

Pope, Alexander. "Eloisa to Abelard." In *The Works of Mr. Alexander Pope*, 391–408. London: Boyer, 1717.

———. "Epitaph Intended for Sir Isaac Newton." In *The Major Works*, edited by Pat Rogers, 242. Oxford: Oxford University Press, 2008.

———. *The Works of Alexander Pope, Esq; Vol III. Part II. Containing the Dunciad, Book IV. And the Memoirs of Scriblerius. Never Before Printed*. London: Dodsley, 1742.

Powell, Manushag N. *Performing Authorship in Eighteenth-Century English Periodicals*. Lewisburg, PA: Bucknell University Press, 2012.

Raven, James. *London Booksellers and American Customers: Transatlantic Literary Community and the Charleston Library Society, 1748–1811*. Columbia: University of South Carolina Press, 2002.

———. *Publishing Business in Eighteenth-Century England*. Woodbridge, UK: Boydell Press, 2014.

Reynolds, Loni. "'The Mad Ones' and the 'Geeks': Cognitive and Physical Disability in the Writing of Jack Kerouac and Allen Ginsberg." *Journal of Literary and Cultural Disability Studies* 9, no. 2 (2015): 153–69.

Rider, Robin. "Perspicuity and Neatness of Expression: Algebra Textbooks in the Early American Republic." *Science in Print: Essays on the Science and Culture of Print*, edited by Rima D. Apple, Gregory J. Downey, and Stephen L. Vaughn, 37–64. Madison: University of Wisconsin Press, 2012.

Robins, Benjamin. *Mathematical Tracts of the Late Benjamin Robins, Esq*. London: J. Nourse, 1761.

Rose, Mark. "The Author as Proprietor: Donaldson v. Becket and the Genealogy of Modern Authorship," *Representations* 23 (Summer 1988): 51–85.

Rosenblatt, Louise. *The Reader, the Text, the Poem: The Transactional Theory of the Literary Work*. Carbondale: Southern Illinois University Press, 1978.

"Samuel Jones." *City Readers, New York Society Library*. http://cityreaders.nysoclib.org/Detail/entities/583.

Scarry, Elaine. *Dreaming by the Book*. Princeton, NJ: Princeton University Press, 2000.

Secord, James A. "Newton in the Nursery: *Tom Telescope* and the Philosophy of Tops and Balls, 1761–1838." *History of Science* 23, no. 2 (1985): 127–51.

s'Gravesande, Willem. *Mathematical Elements of Natural Philosophy, Confirm'd by Experiments: or, An Introduction to Sir Isaac Newton's Philosophy*. London: W. Innys, 1747.

Shadwell, Thomas. *The Virtuoso* (1676). Edited by David Stuart Rhodes and Marjorie Hope Nicolson. Lincoln: University of Nebraska Press, 1966.

Shanahan, John. "Theatrical Space and Scientific Space in Thomas Shadwell's *Virtuoso*." *SEL: Studies in English Literature, 1500–1900* 49, no. 3 (2009): 549–71.

Shank, John Bennett. *The Newton Wars and the Beginning of the French Enlightenment*. Chicago: University of Chicago Press, 2008.

Shapin, Steven. "'The Mind is its own Place': Science and Solitude in Seventeenth-Century England." *Science in Context* 4, no. 1 (1991): 191–218.

———. *A Social History of Truth: Civility and Science in Seventeenth-Century England*. Chicago: University of Chicago Press, 1994.

Shapin, Steven, and Simon Schaffer. *Leviathan and the Air-Pump: Hobbes, Boyle, and the Experimental Life. Including a Translation of Thomas Hobbes, "Dialogus Physicus de Natura Aeris," by Simon Schaffer*. Princeton, NJ: Princeton University Press, 1989.

Sher, Richard. *The Enlightenment and the Book*. Chicago: University of Chicago Press, 2007.

Sherman, William. "What Did Renaissance Readers Write in Their Books?" In *Books and Readers in Early Modern England: Material Studies*, edited by Jennifer Andersen and Elizabeth Sauer, 116–34. Philadelphia: University of Pennsylvania Press, 2002.

"*Sir Isaac Newton's Philosophy.*" *City Readers, New York Society Library.* http://cityreaders.nysoclib.org/Detail/objects/3386.

Slowik, Edward. "Descartes' Physics." In *The Stanford Encyclopedia of Philosophy*, edited by Edward N. Zalta. Stanford University, Fall 2017. https://plato.stanford.edu/archives/fall2017/entries/descartes-physics/.http://plato.stanford.edu/entries/descartes-physics/.

Smart, Christopher. "The Temple of Dulness." *Poems on Several Occasions.* Eighteenth-Century Collections Online: Text Creation Partnership. http://quod.lib.umich.edu/e/ecco/004808051.0001.000/1:11.2?rgn=div2;view=fulltext.

Smith, Courtney Weiss. *Empiricist Devotions: Science, Religion, and Poetry in Early Eighteenth-Century England.* Charlottesville: University of Virginia Press, 2016.

Spacks, Patricia Meyer. *Privacy: Concealing the Eighteenth-Century Self.* Chicago: University of Chicago Press, 2003.

Spedding, Patrick. "Eliza Haywood's Eighteenth-Century Readers in Pennsylvania and New York." *Australian Humanities Review* 56 (2014) http://www.australianhumanitiesreview.org/archive/Issue-May-2014/spedding.html.

St. Clair, William. *The Reading Nation in the Romantic Period.* Cambridge: Cambridge University Press, 2004.

Stewart, Larry. *The Rise of Public Science: Rhetoric, Technology, and Natural Philosophy in Newtonian Britain, 1660–1750.* Cambridge: Cambridge University Press, 1992.

Swann, Julian. *Politics and the Parlement of Paris Under Louis XV, 1754–1774.* Cambridge: Cambridge University Press, 1995.

Swenson, Rivka. "Optics, Gender, and the Eighteenth-Century Gaze: Looking at Eliza Haywood's *Anti-Pamela.*" *Eighteenth Century: Theory and Interpretation* 51, nos. 1–2 (Spring/Summer 2010): 27–43.

Taylor, Derek. "Clarissa Harlowe, Mary Astell, and Elizabeth Carter: John Norris of Bemerton's Female 'Descendants.'" *Eighteenth-Century Fiction* 12, no. 1 (1999): 19–38.

"This Day Is Published by Elliot & Kay." Edinburgh: Elliot and Kay, 1788.

Thomas, Antoine-Léonard. *Eloge de René Descartes.* Yverdon, 1765.

Thomson, James. "A Poem Sacred to the Memory of Sir Isaac Newton." In *The Seasons, 57–69.* London: Millar and Millan, 1730, 295 pp.

———. "A Poem Sacred to the Memory of Sir Isaac Newton." In *The Seasons, 1–15.* London: Millar and Millan, 1730, 348 pp.

———. *The Poetical Works of James Thomson.* Aberdeen, UK: J. Boyle, 1777.

———. *The Poetical Works of James Thomson.* London: C. Cook, 1794.

———. "Receipt for Payment by John Millan for sale of copyright in a series of poems," July 18, 1729. pfMS Eng 1123 (2), Houghton Library.

———. *The Seasons.* London: Millar and Millan, 1735.

Thompson, Helen. *Fictional Matter: Empiricism, Corpuscles, and the Novel.* Philadelphia: University of Pennsylvania Press, 2016.

Towsey, Mark R. M. *Reading the Scottish Enlightenment: Books and Their Readers in Provincial Scotland, 1750–1820.* Leiden: Brill, 2010.

"Valentine Peters." City Readers, *New York Society Library.* http://cityreaders.nysoclib.org/Detail/entities/894.

Valenza, Robin. *Literature, Language, and the Rise of the Intellectual Disciplines in Britain.* Cambridge: Cambridge University Press, 2009.

Vickers, Ilse. *Defoe and the New Sciences.* Cambridge: Cambridge University Press, 2006.

Voltaire, François-Marie Arouet de. *Letters Concerning the English Nation.* London: Davis and Lyon, 1733.

Wall, Wendy. *The Imprint of Gender: Authorship and Publication in the English Renaissance.* Ithaca, NY: Cornell University Press, 1993.

Warner, Michael. *Publics and Counterpublics.* Cambridge, MA: MIT Press, 2002.

Wenger, Etienne, Beverly Trayner, and Maarten de Laat. "Promoting and Assessing Value Creation in Communities and Networks: A Conceptual Framework." *Ruud de Moor Centrum.* Heerlen, NL: Open Universiteit, 2011.

Westfall, Richard S. *Never at Rest: A Life of Isaac Newton.* Cambridge: Cambridge University Press, 1980.

Whiteside, D. T., ed. *The Mathematical Papers of Isaac Newton. Vol. 8, 1697–1722.* Cambridge: Cambridge University Press, 1981.

Wigelsworth, Jeffrey. "Competing to Popularize Newtonian Philosophy: John Theophilus Desaguliers and the Preservation of Reputation." *Isis* 94, no. 3 (2003): 435–55.

Willard, Thomas. "The Metamorphoses of Metals: Ovid and the Alchemists." In *Metamorphosis: The Changing Face of Ovid in Medieval and Early Modern Europe,* edited by Alison Keith and Steven James Rupp, 151–64. Toronto, ON: Centre for Reformation and Renaissance Studies, 2007.

Wilmot, John, Earl of Rochester. *The Farce of Sodom, or The Quintessence of Debauchery.* London: Joseph Streater, 1689.

Wilson, Penelope. "Classical Poetry and the Eighteenth-Century Reader." In *Books and Their Readers in Eighteenth-Century England,* edited by Isabel Rivers, 69–96. New York: St. Martin's Press, 1982.

Woolfe, Richard J. *Marbled Paper: Its History, Techniques, and Patterns.* Philadelphia: University of Pennsylvania Press, 1990.

"Works of Buffon." *City Readers, New York Society Library.* http://cityreaders.nysoclib.org/Detail/objects/1297.

"William Samuel Johnson." *City Readers, New York Society Library.* http://cityreaders.nysoclib.org/Detail/entities/573.

Williams, Julie Hedgepeth. *The Significance of the Printed Word in Early America: Colonists' Thoughts on the Role of the Press.* Westport, CT: Greenwood Press, 1999.

Wordsworth, William. *The Prelude, or, Growth of a Poet's Mind*. London: Edward Moxon, 1850.

Yeo, Richard. *Encyclopaedic Visions: Scientific Dictionaries and Enlightenment Culture*. Cambridge: Cambridge University Press, 2001.

———. *Notebooks, English Virtuosi, and Early Modern Science*. Chicago: University of Chicago Press, 2014.

Index